About this Book

The Berwick community today lingers only in the minds of a few old timers. Otherwise, the community has passed from the scene, vanished as might a whirlwind on a scorched, summertime Ozark hayfield.

Such is the fate, not only of this community, but also of a myriad of other rural communities throughout the land. In their time, these communities fostered a rural way of life, succoring generations of young people and contributing to America's agricultural might. Now, with the changing agricultural scene, many are languishing and some have already disappeared.

The passing of these communities is to be lamented not just because they are nostalgic appendages of the past. These communities produced food both for the nation and the world. They contributed to the stability of the country. A warning of the possible consequences of this loss is contained in these writings.

In doing so, this book traces the birth, life, and now near disappearance of one of America's agricultural communities—the community of Berwick nestled in the Clear Creek Valley in southwest Missouri.

About the Author

Forty years after leaving the family farm, Dr. Robert McGill returned to farm a portion of the family homestead. The family moved to the farm when he was twelve years old. In this book Bob reflects on the joys of life as a youngster, of family and friends and being surrounded by farm animals, and the daily hard work of growing up on a small Ozark farm.

Bob never strayed far from the Ozarks. He attended the University of Missouri and the Union Graduate School (now Union Institute and University) before returning to southwest Missouri where, for seventeen years, he was employed by the University Extension Service. Today, through a not-for-profit corporation, he works with the local Elderhostel program.

After the forty year hiatus, and at the death of his parents, Bob, as a part-time farmer, began to farm a part of the family homestead—this time as the person making the major decisions on the seventy acre farm. Rebuilding the farm, although hard and tedious labor, remained enjoyable work. Here, from the heartland of America, are his views on the changing American agricultural scene.

Appreciation

No one stands alone, and that is particularly true of writers. I gratefully acknowledge that many people, some through their lives and actions and others through direct participation, have contributed to this book.

My sisters Pat and Dott, and my wife Karlene, have, to varying degrees, lived much of this book. Others whose help I appreciate are nephew Bob; friend and copy editor par excellence, Frank Reuter; Ron Macher, founder and publisher of the *Small Farm Today* magazine; and the assistants in the genealogy room of the Newton County Library, Virginia Randolph and Rilla Scott. Marilyn Perlberg contributed an excellent "final read" to the manuscript. Dan Haase kept me grounded in the realities of farming. Others include Basil Ferguson, Mike Ferguson, Willis Finn, Dennis Yoder and Fred Pfister; my college professor from years past, Mary Sellers, who taught me about communities; and fellow writers Ellen Massey, Barbara Wehrman, and Betty Henderson; and neighbors Bill Butterfield and Audrey and Paul Greenquist. And, the wonderful people at Newbattle Abbey College, Dalkeith, Scotland, hosted us during our visits there. Pat Weber and Bill Hunt supported me when I needed help the most. Jeff St. Clair designed the book cover and Kristel Hunt typeset the book.

Return to the Farm

Robert McGill

Library of Congress Cataloging-in-Publication Data

McGill, Robert.
Return to the farm / Robert McGill.-- 1st ed.
p. cm.
Includes bibliographical references (p.).
ISBN 0-935069-06-2 (hardback : alk. paper) --
ISBN 0-935069-07-0 (soft cover : alk. paper)
1. Farm life--Berwick Region--Missouri--Anecdotes.
2. Family farms--Berwick Region--Missouri--Anecdotes.
3. Agriculture--Berwick Region--Missouri--Anecdotes.
4. McGill, Robert. I. Title.
S521.5.M8M36 2005
630'.9778'732--dc22
2004021230

Copyright © 2005 Robert McGill

All rights reserved. No part of this book may be reproduced or transmitted in any form or by any means, electronic or mechanical, including photocopying, recording or by any information storage and retrieval system, without written permission from author, except for the inclusion of brief quotations in a review.

Printed in the U.S.A.

10 9 8 7 6 5 4 3 2 1

Contents

Preface	7
New Pioneers	13
Berwick the Childhood Years	18
The Transition	32
Baby Calf Time	36
Grinding Grain	39
Mother	48
Charley and Me	52
Doc Charley and other Doctors I have known	61
B.F. and Kate's Jersey Cow	79
Pecan Orchard	83
The Year 2000	99
Berwick Community Church	109
Clear Creek	116
Who Owns My Farm?	130
Berwick Return to the Farm	142
Suggested Reading	157

Preface

In the mid-90s after the death of my father and then my mother, I inherited a small seventy acre farm, a part of the family farm where I had grown up. The farm, located a full hour's drive from the home my wife and I shared, became an active pursuit. At least twice a week, I would maneuver one of our family vehicles—either the rust-eaten, rattley old truck or, of more recent vintage, our foreign-made four-wheel-drive station wagon—up and down the valleys and around the curves of the Ozark hillsides, traveling to and from the farm. In driving back and forth, my mind wandered. I found myself thinking about returning to the family farm I had grown up on forty years earlier and all the changes that had been wrought in my lifetime. The realization was inescapable. A significant part of my life was tied to a piece of Ozark bottomland. Thoughts about the importance of this piece of land spontaneously took center stage in my brain.

I wondered if it would be possible to write a book about these reflections. Soon, whatever welled up within my mind about the farm during these drives became the raw material for the book. Events from my youth and the tales of old-timers—memorable stories about farming with cantankerous old mules, fishing down by the creek with a cane pole and worms, or hunting raccoons and possums by night with baying hound dogs—still made me laugh. In succession I remembered a painful broken arm sustained one summer while chasing a wayward cow, riding the yellow school bus to and from school, warding off stage fright at the yearly Christmas pageant at the church across the road from our farm, and tending to our family's herd of dairy cattle. Even recent events became suitable material for the book, for, as I thought about my fledgling pecan orchard, I realized that I had always loved walking in the woods and looking up into trees, although as a kid it was usually done with a squirrel rifle in my hand. Rebuilding the farm fences reminded me of my father's old-time work ethics.

On arriving home at the end of each trip to the farm, I would rush into the house and, with the hope of capturing these ideas before they disappeared from my mind, fling words into my Macintosh; misspelled words, partial sentences, and convoluted paragraphs all went into the computer.

Early on, it became evident that the story was not only about our family and growing up on a small farm, but that we were bound together with our friends and neighbors in our community, Berwick. In fact, it was the Berwick community that gave meaning to our lives through school activities, church services, and community "get-togethers." It is impossible to write about the farm without writing about the community.

My desire is to describe, to freeze into place as one might in a portrait, the important features of the community as they existed, both then and now. And, of course, since a community is always changing, to capture the passing joys, dreams, and passions—as well as the tensions and conflicts inherent in community and family life. The lifeblood of the Berwick community seemed to flow on forever, like the water in the small stream that runs through its valley, emerging from a source several miles above the community and continuing on—if one cared to track the total course—to the depths of the oceans.

To understand and clarify my insight into farm life, it was also necessary to reflect on myself and how I had changed in the intervening years. Normally, it was fun to think about growing up on the farm and attending community events with my sisters, Pat and Dott, old friends, and boyhood chums. Life was as full of passion then as it is now. But there were also moments when I remembered the untoward actions of my youth, the times I was unfair to or overly demanding of others.

These, the warts and blemishes of my life, also surfaced. I needed to learn about myself, for I could only explain with clarity as I understood myself. In fifty years much had changed about myself, the community, and my appreciation of the community.

While writing, I was reminded again of what I had learned in college many years before: that many of youth's lessons are hidden deep in the recesses of the mind and sometimes appear only unexpectedly. One of the greatest surprises I encountered was to discover the permeating influence of religion on my life, an importance that I was not conscious of and was surprised to uncover. The theme surfaced time and again. Yes, my mother was devout in her religious life and raised her three children to be equally diligent in our religious practice, but that was many years ago. Since then, although my wife and I attend church with regularity, I had thought little about the connection between my religious upbringing and present values. What I learned as a child continues to be although perhaps in some unorthodox ways a beacon for my life and thus is reflected in this book.

Another of life's great lessons was thrust on me as I was writing the chapter comparing medical treatment of yesterday and today, explaining how current medical treatment now may be more advanced technologically, but certainly does not have the personal touch of medicine fifty years ago.

While in the actual process of writing the article, I was diagnosed with a life-threatening immune disorder, the first time in my life that I had to confront, so starkly, my own mortality. Much of the article deals with the impersonal nature of current medical care. When this chapter of the book was nearly completed, I gave it to my long-time doctor, an excellent physician whom I had known for years. It was not, I'm sure in retrospect, a good move to give an article with a biting edge about the medical profession to one of the medical doctors responsible for diagnosing and treating a life-threatening disorder. After all, I can still remember the agitated look on his face when he had to tell me, forthrightly, the nature and possible consequences of the immune disorder. Fortunately, he was more understanding of me than I was of the impersonal nature of medical care, and he and his colleagues helped me pass through this experience with little more than a few small surgical scars. The experience added an immediacy to life that I did not have. I gained an even greater appreciation for the dearness of enduring personal relationships which have lasted a lifetime.

This book reflects on my experiences from youth to the present, and from farm life to what, again, is farm life. My childhood on the farm, like the childhood of so many others everywhere, was a time of joy and happiness and creative wonderment. Later, as an adult, I learned that childhood is important not just for the memories, but because this is the period of life when the values, beliefs, and customs that sustain a person over a lifetime are formed, ever so spontaneously, and then passed from one generation to another. It's the "up-bringing" or character development that my parents worried so much about that was important. And normally, this learning about how to deal with life's most important questions comes from family life and community activities, even though it may be an unconscious process. Kids are too busy being kids to really appreciate all that is happening to them. Usually we do not even recognize the importance of childhood until we reflect on the period—perhaps not until fifty years later.

So, a significant portion of this book treats the values of a rural community, in this case the Berwick community, which even now, has nearly lost its identity. But, if one learns to identify values, one can place value judgments. And, surely, if my generation has learned anything, it is that bigger is not always better, nor that more sophisticated is somehow superior, nor that the more intricate and marvelous our technology, the more stable and mutually supportive our society becomes. We experience problems by relying on bureaucracies, technology, and our bigness—and some of these difficulties, if not properly addressed, could eventually result in our downfall.

Intrinsic to a democratic society is the obligation to continually examine existing practices and courses of action. Alternatives for the future would always seem to be possible and helpful, especially when dealing with healthy personal relationships, families, communities, and, if you will, the nation. In this book the focus is on agriculture. I am convinced—though I don't want to become too preachy on the subject—that as a nation we can have a better and more healthful food supply for our citizens. A more prosperous agricultural economy, especially for family farming, is possible. A better road map for agriculture exists. Parts of this map are tied to the past, like the small farm I grew up on and still maintain.

Unfortunately, governmental support of farming, although massive in terms of dollars, does little to help the viability of small agricultural entities. The preponderance of the huge payments goes to large corporate farms while the smaller farms, those normally tended by families, receive little help. The unintended consequences are that the government actually supports the removal of families from farms. Unintended governmental consequences abound. Some nations plant land mines to deter their enemies, and then their own children step on the mines and blow their legs off. In this country, in agriculture, we subsidize the huge farms to promulgate farm efficiency, and inadvertently run highly skilled farmers off their land and into less-satisfying occupations. Not only do these transformed farmers suffer, but their own communities—and eventually the nation as a whole—feel the effects of their removal.

Fortunately, in this nation, and in farm life in particular, the independent spirit is indomitable and individual initiative flourishes, fostered, I believe, by farm life. Many people, some well-established family farmers and others, like myself, novices tending the land, enjoy the struggle to create a viable farm. To do so, some of us travel some rather unpredictable avenues. And, a very few of us write books about our efforts to do so. But rarely, I think, is it possible to catch glimpses of agricultural community life over a period of fifty years— which is roughly the length of time I have been associated with the Berwick community. I am, to say the least, highly biased in my opinion of the community. I have many fond memories of growing up there; my childhood recollections of my parents and sisters remain tied to events in the community, and many of my values, opinions, and beliefs were formed in this community. And, by happenstance, I still make weekly visits to the farm. This community is as much a part of my being as the food I eat, the air I breathe, and the water I drink. It is impossible to separate my personal experiences from the lessons I learned. Thus, I present them both, small sketches of my life and suggestions of how our agricultural society might become fuller, richer, and more life-sustaining.

New Pioneers

I remember the day over fifty years ago when, as a boy, my father took me on a return trip to our recently vacated home in Enid, Oklahoma. It was a special treat for me because just he and I went, leaving my mother and two sisters alone for a couple of days to tend to our new farm. In Enid, we loaded the last of the family's belongings on a pickup truck that was old, even then, and made the final moving trek to the Ozarks. The overloaded pickup with cartons, crates, and furniture bulging from the stock racks, was reminiscent of the mythical Beverly Hillbillies' move to Hollywood which would take place years later.

And yet my father, as did my mother and two sisters, hoped for a different and better life on an Ozark farm. It's not that life was bad in Oklahoma. My father, a World War II veteran, was well paid as a union electrician, and our family was stable. He simply wanted to become a farmer, and Mother agreed to the move.

Our 130-acre farm, like other farms in the area, was diversified—that is, Dad raised a variety of crops and kept several kinds of farm animals. Normally, Dad raised wheat, oats, barley, and corn and, although relying on milk cows for most of the farm income, he also kept sheep, hogs, chickens, guineas, and geese. The diversity of the farm kept us busy, but as Dad often told us, "We always have something to sell if money gets a little tight."

The first year on the farm was especially exciting. The transfer from hand labor to mechanization was nearing an end in the Ozarks at that time. That first year was the final year the neighborhood ran a threshing crew. And so I had the opportunity to bind grain and feed it into a monstrous, bellowing threshing machine. For several months we also milked our first thirteen dairy cows by hand, twice a day, before purchasing a

Return to the Farm

milking machine. I remember that I opposed the purchase of the milking machine as an unnecessary expense. And yet, as the dairy herd grew, I became aware of its value, and my unfounded opposition.

Farming, at least at first, was not easy for my parents. In 1952 and 1953, great droughts came to the Ozarks, and we we were forced to cut down red oak trees so that the hungry milk cows could eat the wilted leaves. The cows didn't give much milk from eating the red oak leaves, but they survived.

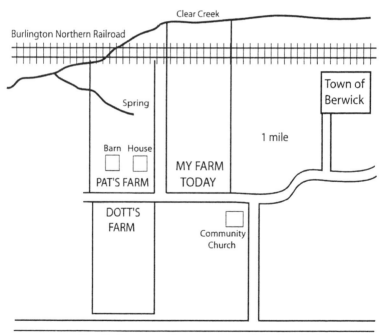

Berwick Community and Our Farm
1950 & 2000

Originally all three parcels were a part of the family farm.
The parcels are approximately seventy acres each.

New Pioneers

Our family survived in agriculture, too, only because my parents periodically worked "off the farm." Dad kept a round-trip ticket to his union job back in Oklahoma, and on several occasions he absented himself from the farm for months at a time in order to work at his old craft there—or he worked in several of the larger towns of the Ozarks. Mother, too, at various times worked in the local garment factory, at a casket factory, and as a bookkeeper. Their employment supplemented the income from the farm and allowed us to remain a farm family.

The farm was constantly evolving. As the years passed, we became more productive farmers, and the farm grew to 200 acres. Slowly, Dad reduced the number of field crops until he raised only corn and hay. One year we did not replenish the chicken flock and tore down the chicken house. When I graduated from high school and left, my father sold the last of the hogs. When the formal education of my younger sister was complete, he converted the farm from a dairy to a beef operation.

From then on Dad kept beef cattle, and he and Mom lived on the farm until they both died. At one time, toward the end of their lives, they talked about leaving the farm and moving to town, but decided they did not want to live in even a small town. Farming had become too much a part of their being. It was all they knew. In the end it was all that satisfied them.

Life was never particularly easy for them. After they moved to the farm, they could never afford a new car. However, they helped each of their children receive a good education. Both parents never forgot the names of each of the many family pets, in succession, over the years. Dad could recite the name and production record of each milk cow he ever owned. When my parents got sick or left on short trips,

Return to the Farm

the neighbors gladly volunteered to do their chores for them. Ozark agriculture was a way of life for my parents.

Today, there is a new, revitalizing trend in agriculture. It began several years ago when a few, mostly new small landowners, began talking about establishing a variety of commercial, often non-traditional, enterprises on one farm, instead of specializing in one crop. They began raising goats and rabbits, and wondering about the profitability of catfish, crawdads and frogs, as well as beef and dairy cows. They grew and marketed the produce from small patches: vegetables, tomatoes, strawberries, blackberries, luscious blueberries, and fancy mushrooms. Some of these farmers began making cheese. Others sold beef, goats, and sheep directly to consumers. Eventually, a few established farmers began to supplement their incomes with these new enterprises. Today a trend has developed. Many more farmers are developing diversified farms, much as my father did fifty years ago. The trend, I think, continues to grow. Whenever these farmers gather, they talk not only about production, but also about the marketing of the new enterprises. The trend has taken on the common name of alternative agriculture.

I think it is not too far-fetched to compare the tenacity and vision of those now engaged in alternative agriculture with that exhibited by my father and others of a generation ago. The farmers of a generation ago did not know the direction agriculture would take, nor did they find any easy solutions. Nevertheless, they helped create a system that sustained them over a lifetime.

Current farmers in alternative agriculture undoubtedly need to sort and cull among the various possibilities and experiment with their chosen enterprises. They will make many mistakes, but the same promise exists for them that existed a generation ago. I suspect many will be successful, not only in

New Pioneers

their individual efforts, but also in building a new agriculture that does more than attempt to perpetuate the past. Their influence will be disproportionate to their numbers, and their contributions to a new rural economy will endure.

Note

A version of this chapter appeared in Volume 5, Number 2 issue of *The Missouri Farm* magazine, (now *Small Farm Today*), 1988.

Berwick
The Childhood Years

Nothing could have been more picturesque or promised more adventure than the farm Dad purchased in 1952. To begin with, Mother, although she questioned the move to the farm, believed the farmhouse was perfect for raising her children. The imposing farmhouse, located at the top of the hill and at the intersection of two country roads, could be seen well in advance by anyone approaching. In return, approaching vehicles could be seen long before they got to the house. The farmhouse, blending in perfect symmetry with the stately maple and wild cherry trees that surrounded the house, graced the hilltop. By its bearing alone, our new home offered a gentle welcome to all who approached.

A huge porch wrapped two sides of the house, a great place, we would learn, for resting at the end of a long day or for looking at baseball cards on rainy days. An old smokehouse attached to the house, although converted years earlier to a storage room, still retained, embedded in the timbers of the ceiling, the aroma of hickory-smoked hams and bacon. On still days this faint but delightful smell enveloped the whole house.

The large downstairs consisted of only three rooms: a kitchen, living room, and dining room. The hardwood floors of each room sparkled and added a sense of warmth to the house. The kitchen, where mother cooked and canned food, immediately became the focal point of family life. It was here that much of the family conversation took place. We ate meals in the dining room, beside the oft-lit fireplace, while the living room, often hidden behind closed folding doors, was used for formal occasions, like entertaining guests after Sunday dinner.

The upstairs contained four unheated bedrooms, one for each of us children and one for my parents. In the wintertime we snuggled under voluminous down ticks and quilts to keep warm.

 A magnificent view of the bottomlands presented itself from the upstairs back window. For from there we could check on cattle at the far end of the pasture, catch the first glimpse of gathering storm clouds, or watch the overhead passage of ducks and geese at the change of the seasons. From here, too, we saw pencil-thin trains threading their way along the railroad tracks, and witnessed the dramatic overnight change of the engines, from puffing steam engines to diesel locomotives.

 Meanwhile, the stairway was a great place to play and the bannister to slide down, when Mother was not looking.

 The house, thankfully, had indoor plumbing and Dad installed a big, black pot-bellied stove, trimmed with chrome, for wood heat. Since the house was so large and without insulation, the stove burned prodigious amounts of wood and I would soon learn to cut firewood, first at one end of a cross-cut saw and later with a chainsaw.

 Could my parents have selected any better place to live? Memories of growing up in the house—where we three children ate, slept, played together, received love and warmth and tutelage from our parents, and invited friends until the time we each departed from home—will remain forever. And, the house served my parents long after we were gone, for our parents enjoyed a great rapport with many people throughout their lives and invited a wide variety of friends and neighbors into this home for as long as they lived.

 But to a kid fresh to the farm, the barn offered the greatest adventure. The massive structure, constructed years earlier from huge native timbers by workmen using primitive hand tools, invited activity. At some time in the past workmen had added twelve stanchions along the west side of the barn. Cows, I learned immediately, have their own pecking order

Return to the Farm

and at milking time our dairy cows, in expectation of being fed and milked, would charge, twice daily, into the barn to their own stanchion in the exact same order. Learning to milk a cow—a necessary chore repeated both morning and night—became my introduction to farm work.

Next to the milking shed, but on the south side of the barn, Dad kept gunny sacks full of dairy feed for the cows. Dad knew the amount of milk each cow gave, and fed them accordingly. Big, heavy-producing cows received several scoops of dairy feed at each milking, the smaller cows, further into their lactation, much less. Feed was expensive, and adjusting the amount given to each cow controlled costs. Meanwhile, farm cats loitered around the feed area, hoping to catch the occasional mouse that would run from beneath a feed sack. And, at milk time, the cats would saunter into the milk parlor and sit on their haunches, tail curled back around their bodies, expectantly waiting for the inevitable stream of milk we learned to squeeze with precision from the cow, pinpointing the warm flow directly into the cats' open mouths.

Beyond the milk parlor were small pens, nurseries for the baby calves. One of my first jobs was to teach baby calves to drink milk from a bucket. I quickly learned that even three-day-old baby calves exhibited uncanny strength, resisted being held, and hurt me with head butts.

I would pour a quart of milk into a bucket, enter the small calf pen quickly but quietly, straddle, and then back one

Feed Storage	Milking Parlor	Grain Storage
Calf Pens	Hay Storage	Grain Storage

The Barn

of these unsuspecting but wily creatures into a corner before it realized what was happening. Next, I'd place my two middle fingers deftly into the calf's mouth so as not to get my fingers bitten by the calf's sharp and powerful teeth, cram the baby calf's head down to the bottom of the bucket, and listen to the calf spew and gargle until it got a first taste of milk. An unsuspecting visitor might have thought I was trying to drown the poor baby calf, but I would always soon feel the rhythmic tug and pull on my fingers as the calf began to suck, a sign the calf was beginning to drink. Finally, quickly but carefully, I'd remove my fingers from the sucking calf's mouth and the calf would be on its own, drinking milk. The key to the process was to be firm. Coddle the poor little babies and I'd be turned upside down in a smelly calf pen and tromped on by the sharp hooves of a precocious baby calf while my clothes would be soaked with spilled milk.

Two granaries covered the entire north side of the barn. The smaller of the two granaries held field grain, wheat, oats and barley. It was in this granary that I learned the back-breaking job of scooping grain. "You'll have to find a scoop that will fit your hands," the neighbors used to tease me and laugh when they watched me scoop grain. Dad bought a big scoop shovel, one that would slice through the grain when we filled the granary in the summer, then slice again in the wintertime when we emptied the bin. The second granary, much larger than the first, held ear corn. This granary had slated sides so that air could move through the corn, drying it even more after harvest. Because of corn's irregular shape, I would learn that eared corn was much more difficult to scoop than the small field grains.

The really fun part of the barn, though, was the second level, the hay loft, where we stored baled hay. Baling required exhausting work, always on the hottest days of summer. Dad

Return to the Farm

would cut the hay, let it lie in the field to dry underneath the beating sun for a day or two, then call for the owner of a hay baler to come and bale the cured hay. Finally, the hay crew would arrive to put the hay in the barn. Oh, what excitement that was! The hay crew, high school boys just older than me, would "buck" the hay onto hay wagons in the field, haul them to the house, and unload the bales, forming a human chain and throwing them, one to another, until they had packed the barn, clear to the roof, with bales of sweet-smelling hay. I anticipated the time, just a few years in the future, when I would be big and strong enough to work on a hay crew.

    The hay loft was the perfect place for my sister, Pat, and me to play, either by ourselves or with friends, on balmy afternoons or rainy days. It was here, in the hay loft, that we built forts, played doctor, and talked about what we would do when we became adults. Occasionally, too, as we whiled away our time in the loft, a black snake would slither past, causing shrieks of panic from the girls and brave actions from us boys.

    Surprises even awaited us in the garden space, located a few hundred yards below the house, at the bottom of the hill. A nearby spring emerged near the garden and flowed the length of the farm. As we tended to the green beans, picked peas and/or dug potatoes, we would, occasionally, find an Indian arrowhead. Obviously, at some time in the past, an Indian village had been located around the spring, on the very spot where our garden was located. After a find, we could conjure up stories about Indians and that arrowhead for half a day.

    The spring branch itself, just barely wide enough for us to jump across if we got a good running start, flowed crystal clear—the water pure enough at that time to drink directly as it emerged from the ground. The stream, lined with watercress, glistened in the sunlight as it meandered through the pasture while schools of minnows darted past and crayfish, half buried in the mud, hid on the bottom. In our exploration, Pat and I were obliged to cut our way through the huge weeds that grew on the banks of the spring branch, a task that made the rewards for whatever we might find—an arrowhead, animal bone, or feather from a hawk—all the more rewarding. We earned whatever we found. Lewis and Clark could have sensed no greater feeling of achievement when they discovered the confluence of the Missouri and Yellowstone Rivers than did we kids the day we found the spot where our spring branch entered into Clear Creek, a few hundred yards below our farm.

One of Dad's first purchases, other than the cows, was the half-grown hog he planned to keep as the family brood sow. The young gilt, named Flossie, was born a runt but adopted and raised as a pet by neighbors who could no longer care for the growing animal. They sold it to Dad rather than through a livestock auction barn—a sure sign Flossie would have been an endangered animal, headed for the slaughter house.

Dad built a small pen for Flossie where we could tend to her, and she could continue to grow and become a mother. But Flossie had other ideas about where she might stay. Flossie discovered early on that wherever she could get her snout through a fence, she could also, eventually, by pushing and shoving, get her body through. And, because Flossie had been raised a pet, she became lonesome in a pig pen out in the barn lot, away from human beings. So, from time to time we'd see Flossie, having freed herself from her pen, ambling to the house where she would lie down in Mother's flowerbed, waiting for us to come out and rub her belly with our doubled-up fist and gently drive her back to her repaired pen—hoping this time she would stay. Even several years later, after she became a brood sow, she would still, on occasion, escape from her pen and, with a litter of squealing piglets running behind her, waddle up to the front yard and again flop over into Mother's flowerbed, waiting for our attention.

The most fertile land on the farm stretched a half mile across the bottom, from the foot of the hill below the house to Clear Creek. Fertile soil it was, too, and it would grow about anything Dad wanted to plant on it. Dad reserved this land for field crops of oats, corn, barley, and wheat, and it was on this land that I learned to drive our small Ford tractor, plowing, preparing a seed bed, and then drilling wheat or planting and cultivating corn and cutting hay. Here, too, grew the lush fields of legumes that we mowed and baled as hay and hauled to the barn on hot summer afternoons.

Clear Creek, a small, spring-fed stream containing an abundance of fish and frogs and crawdads and other aquatic life, and with pools just deep enough for swimming, formed the northern boundary of the farm. Trees lined the banks of Clear Creek and small paths, made by wildlife creatures, coiled through the nearby underbrush. Who knew what animals might lurk near the creek? Clear Creek was a perfect place for family recreation or children's adventurous sorties.

It is no wonder that Dad, from the moment he purchased the farm until his death, was proud of this farm. Instantly, he became devoted to it. We children gave little thought to what it immediately became apparent to our parents: that the farm would not provide an adequate income to supply our family's modest needs and pay the mortgage, too. Within months of moving to the farm, Dad was required to go to work "off the farm" just to "make ends meet." Thus for several years Dad was an electrician by day and a farmer by night and on weekends, with the rest of the family helping with the abundance of farm chores, all in order to "make a go" of the farm. Yet, to my knowledge, my parents never seriously considered selling the farm and returning to Oklahoma. Together, they had

Berwick The Childhood Years

decided on a farm life. We children gave little thought to the fact that life could have been different. We just accepted the decision Mother and Dad had made. The farm quickly became our childhood home.

By virtue of the move, however, we more than moved onto a farm and into a farmhouse. We also found ourselves in an old established Ozark community, the community of Berwick. Other small towns and communities dotted the countryside: Needmore, Smack-out, Wentworth, Ritchey, and the Fox community, each named after a particular event, family, or physical feature. Berwick was named after a small railroad siding that once existed about a mile and a half below our house. We learned very quickly that our new neighbors had a strong attachment to the Berwick community and that they believed all who lived in the community were among the salt of the earth. All were God-fearing, hard-working, frugal people who looked after and shared with one another. Thus, it was with some trepidation that Mother and Dad, hoping that our city ways would not be too foreign to our neighbors until we learned about farm life, set about to meet our neighbors. They wanted to "measure up" to these high standards.

They need not have worried. The neighbors took a liking to our family. Two older couples in the neighborhood, Rose and George and Pat and Leota, folks filled with knowledge about Ozark farm life, took turns hosting occasional Sunday dinners with our family.

These Sunday dinners, some of my first recollections of farm life, remain among my fondest. First, there was the food itself. The dinner table overflowed with farm bounty—peas, beans, fried chicken, roast beef, salad, and strawberries and cake—farm fresh produce all of it and, although we didn't think about it, grown with natural fertilizers, making the dinner an organic feast.

Return to the Farm

But, even stronger in my recollection are the conversations that occurred after the meal on these Sunday afternoons. The women would gather in the kitchen to do dishes and talk about family, friends, and activities; the men would retire to the living room to discuss farming and tell stories. Mixed in with their banter of farming and weather and markets were stories of the past—thrown in, sometimes, I am sure, especially for my benefit.

Our neighbor Pat, in the best tradition of the Ozarks, was undoubtedly the finest storyteller for miles around. Pat was short and plump—not fat—just plump and nearly bald and always smiling.

Pat was missing three fingers on his left hand—only his thumb and forefinger remained. The missing fingers were not too obvious and did not seriously impede his actions. Pat saw me watching his missing fingers one time, with a little horror, I suspect, and showed me the stumps.

"And how did you lose them?" I asked, hesitantly.

"When I was a small boy," Pat responded, "I used to tease my older brothers. One of my brothers, Shock, was chopping wood. To tease him, I put my little finger on the chopping block just before he chopped and dared him to cut it off," Pat said. "Shock took the dare. He cut my finger off."

Pat then told how his Uncle took a straight razor and cleaned off the dangling meat and skin, lapped the remaining skin over the wound and placed a band-aid over it so that in a few weeks the wound healed.

"But I was a slow learner," Pat laughed.

"When the wound healed, I went back out and watched Shock cut more firewood. Just to tease him again, I placed two of my good fingers on the chopping block and dared him to cut 'em off. Sure enough, he cut more of them off too," Pat told me.

Berwick The Childhood Years

"We went through the whole process again. My Uncle got his straight razor out, cleaned off the bone and skin, left a little skin to make a small pad, pulled the skin over the wound, and stuck a bandage over the wound. It healed, but I never again put my finger on the chopping block," Pat said, as he playfully picked his nose with what was left of one of his cut-off fingers. Pat retained the rest of his fingers for the remainder of his life.

Pat also liked to tell stories about working with mules. As everyone in the Ozarks knew, Ozark mules were known to be especially stubborn and cantankerous, even when loyal and prodigious workers. Pat particularly enjoyed telling stories about the time he used his teams of mules to cut hay on hot summer days. He related how bumblebees like to burrow their nests in fields of red clover and, although he would keep a sharp eye out for them, Pat told how he would occasionally miss seeing a nest until he had run over it with the mules and mower. Invariably, the disruption angered the bees who would attack the first moving thing they saw—Pat and his mules. Then, the excitement started. The mules would immediately stampede and start for the house. Pat, in spite of his effort to control the mules which were getting walloped by the bumblebees, could never get the mules in check, and they always broke their harness, leaving the old mower somewhere between the field and house. Pat would have to trudge up the hill, still swatting at the bumblebees, where, eventually, he would find his team of mules in the barnlot near the watering trough, shivering and recuperating from the stings. After calming his mules Pat would apply the same salve to his own stings that he had rubbed on the mules, get a drink, repair the harness and, eventually, persuade the by-now skittish mules to return to the field.

George would respond with stories from his own childhood. He, too, had bumblebee stories. George and his boyhood

Return to the Farm

friends used to play a game with bumblebees, although it took at least two young lads to play the game. First, the boys would make two good, strong, four-foot-long switches from buckbrush plants, switches with good heads on them that could be swished through the air. With these, the boys would set out to find a good healthy bumblebee nest, one with lots of bumblebees darting to and fro, into and out of the ground. One lad would station himself over the hole and begin batting bumblebees as they emerged from the hole—the other would watch for bees honing in from a far distance and beat at these returning bees. If the lads could kill all the bees before getting stung, they won, if not....

There were other stories, too, about hunting and fishing and good quail dogs. George even recalled seeing, in his youth, a family of settlers passing through the region in an oxen-pulled covered wagon on its way to Indian territory. George lived well into his nineties—long enough to also see a man on the moon.

Back then, these Sunday afternoon discussions would often end with the statement that "the good old days were probably better, but I sure wouldn't want to go back there to live."

Soon Dad was "trading" work with his neighbors—a sure sign we had been accepted into the community—at first picking corn, putting up hay, and making silage. It was a tradition he participated in all of his life.

Other community activities I remember include a community sponsored 4-H club. It was a classic 4-H club, with all the kids in the neighborhood, as many as 30 kids of all ages, participating and with monthly club meetings held on a rotating basis in members' homes. Club members would meet in one room with the club leader while all the other parents would socialize in another—the meetings becoming community social events. The older kids were elected officers while the younger

Berwick The Childhood Years

kids did their best to make decisions by the rules of parliamentary procedure. We would talk about our projects: caring for animals for the boys, and sewing and home furnishings for the girls. Meetings normally climaxed with a guest speaker from the county health department, the Missouri Conservation Commission, or a county governmental office, then ended with refreshments—and folks said good-bye until the meeting the following month.

Our school was located outside the community, about eight miles away—a small, two-story, consolidated school where all twelve grades were housed in the same brick building. We lived at the end of the school bus line and so were the first picked up in the mornings and the last dropped off at night. A big yellow school bus (unheated back in those days) would pick us up, normally before sunrise, and we would make the trek to school, greeting other kids as they got on the bus each morning, trying not to play too loud, and arriving at school fresh for another day. I remember Dad used to call the big yellow bus the "Yellow Monster." We would watch for it coming across the hill and always had a wild scramble at the last minute to be ready for the bus when it arrived.

There were twenty-four students in my freshman high school class and at our orientation one teacher shocked us by stating that the normal school drop-out rate was fifty percent. "Only half of you will be here in four years," he said, rather matter-of-factly. "Just be sure you are one of the graduates," he concluded.

He was correct, of course. Our class graduated twelve students.

Religion for us kids was strictly "Baptist." Mother wanted us to understand there was only one way to salvation, and that was through the Baptist church. I was never clear in my youth if anyone other than a Baptist would ever reach heaven. However, because of Mother's instruction and our commitment to the Baptist church, I was certain I would eventually get to heaven. There was no need to question further.

There was a church a quarter of a mile from our house, the Cumberland Presbyterian Church which many of our neighbors attended. But, since it was not a Baptist church, it was, for the first few years after we moved to the farm, unsuitable for us. We attended various Baptist churches in nearby small towns, while occasionally visiting the Presbyterian church. However, to my mother's surprise, services at the Presbyterian church bore a remarkable resemblance to Baptist services, and the teachings were also much the same. In addition, the Presbyterian church served as a social center to the community with periodic Sunday after-church "dinners on the ground" that featured huge amounts of food, followed by song services, and a yearly Christmas pageant put on by church members that attracted nearly every resident of the community. The lure of the activities and the convenience of the church was too great for Mother. She finally decided that we could, after all, attend the Presbyterian church on a regular basis, and she and Father continued to do so for the rest of their lives.

The remains of the old town of Berwick, located about a mile below and east of our house, was once a railroad siding situated next to a spring. Only two buildings remained when we moved to the farm: a small, neat bungalow and a long, dilapidated wood frame shell of a building that years ago had served as a tomato canning factory. Our neighbors, the occupants of the bungalow, lost control of a brush fire one day and inadvertently allowed the ramshackled canning factory building to burn, leaving only the bungalow as a reminder of the town.

But it was not the remembrance of the town that permeated my thoughts, but rather, the community that took its

Berwick The Childhood Years

name from the town. Ask me at the time where I was from, and I would unhesitatingly answer, "Berwick," just as I knew other people from nearby towns who would answer with the names of their towns, or villages, or communities. For me, Berwick was an idyllic place to spend my childhood, for here I learned to be industrious, to feed calves and drive a tractor and put up hay, to interact with neighbors, to devise childhood explorations at the creek with my sisters, and to receive nurture from loving parents. What more could any childhood hold? No wonder it became a part of me.

And then, eventually but abruptly, my life changed dramatically when I graduated from high school. At that time I followed in my sister Dott's footsteps and enrolled in the University of Missouri at Columbia, a town situated in the middle of the state. I can remember the drive to Columbia, then a six-hour drive, over curved roads and through the middle of every small town on the highway. My sister had found an immaculately kept room near campus at Mrs. Jackson's, and it was here that my parents helped me carry my worldly belongings up the stairs in a suitcase. I kissed my mother, who had tears in her eyes, good-bye and shook hands with and hugged my father, and then watched and waved as the family car backed out Mrs. Jackson's driveway, turned, and disappeared up the street. Dad had to get home for a late night milking of the cows. For me, it was the demarcation of my life. Expectations lay ahead. But Berwick, the community of my childhood, was a part of my past, albeit by only a day.

Although I often visited my parents on the farm in the succeeding years, I had no farm responsibilities and only seldom participated in any of the farm chores. For all practical purposes, for forty years, I was in no way associated with the farming operation—until, that fateful time when my mother died.

The Transition

My reintroduction to farming came through the rebuilding of fences on the farm, for they were run down. Over the years the old barbed wire had become brittle and rusty and had snapped, often under the pressure of a breachy cow reaching through the fence for an extra morsel of grass. The old wire had been patched and restrung many times. The old wooden posts leaned over, sometimes almost to the ground, because the old cows repeatedly reached through the fence. If the ground was soft from rain, the posts were pushed from their seating. Slowly but inevitably over the years, they were pushed over until what at one time was a straight line of posts eventually zigged and zagged until it could hardly keep a cow in. Given the fact that Dad, because of his poor physical health, had for years not been able to make more than the most necessary of repairs, I was fortunate to keep part of the old family herd for, as one neighbor said, "Those old cows know where the fence ought to be so they won't get out very much."

I am trying, now, to replace the fence. I want a twenty-year fence. I figure, at the most, I'll farm for twenty years, so I won't have to go back out and do this job again. Slowly, the old fence is being transformed into one that has heavy corner posts concreted into the ground and braced to anchor posts, followed by a long line of straight, evenly spaced new steel T-posts. It's quite a task: tearing out what remains of the old fence, cleaning the brush out of the fence rows, setting the corner posts and driving the new freshly painted metal T-posts, stringing the new wire with the nasty barbs that cut on contact, and finally, fastening the wire to the posts.

It's the second time in my life I've built this particular fence. The first time I was a kid, helping my father. We had just moved to the farm and the fences were run down then, too, so we had to rebuild them almost from scratch. We cut the new

The Transition

posts from hedge trees, pulled out the old wire, and dug holes through the dirt and rocks with a set of hand diggers and pry bar that fit Dad's hands so naturally. Finally, we'd set the posts, tamp them in, and fasten the wire to the posts. "These hedge posts," Dad used to tell me, "are so hard they will outlast their holes." When we first moved to the farm, Dad would work at building fence "night and day," stopping only to do chores and milk the cows. But he didn't mind, he was fulfilling his dream, a veteran returned from World War II.

Mother was always the staunch one in the family, guiding the day-to-day activities of the family. We always knew where she stood on almost any issue, on religion and especially on how we children should conduct our lives. Our father, on the other hand, was the quiet one, laboring daily on the farm for the remainder of his life but rarely interjecting his opinions about raising children. We learned from what he did, not from what he said. The farm and my mother were the only things he seemed to be passionate about, and yet he was absolutely committed to both, and in so doing made a statement which transcended his stoic life.

He left the farm only for short vacations. He wanted to remain close to what was familiar: caring for the family pets, inspecting his herd of cattle every day from the front seat of his dirty-yellow Ford pickup, and conversing with my mother. When he was succumbing, after a long battle with lung cancer and went to the hospital for what we all knew was the last time, he stood gazing out the window a long time at his herd of Hereford cattle before finally being allowed to be escorted, oxygen tank in hand, to the family car.

A handful of years later my mother also passed on and was laid to rest beside my father in the cemetery at the community church just up the hill from our home. At her passing we three children became orphans.

Return to the Farm

Adults rarely talk about becoming orphans. That's a term normally left for young, parentless children. Yet it's equally true of adults, who over the years have come to feel that parents are as much a part of their lives as air and water and good conversation. Eventually, that which has always been is gone. And children discover that, now, there is no generation ahead of them in their family. For so many of us, it is a chilling event which forces us to view life from a new perspective. We were in our fifties when we became orphaned.

So it was for the three of us. The family farm was divided among us. I decided to actively farm my portion and, while the cows normally remember where the fence ought to be, they sometimes forgot. "Your cows are out," is a refrain I heard more often than I wanted, and while I am a part-time, novice farmer, I confess that most of my learning came fifty years ago when I first helped my father on the farm.

Normally, I'm free in the winter months to work on the farm, to clean brush, plant trees, care for cattle, and build fence. Since fencing is my first concern, I concentrate on building the perimeter fences first, then the cross fencing, just as my father did a half century ago. The work is always harder and progresses slower than I plan. There's little use to set a timetable for completing a section since I find it impossible to estimate the length of time needed to complete a task. I just work, sometimes for days, completing one section and, when it's finished, moving on to the next.

Nor can I concentrate solely on the fence. There are other activities that need to be done on the farm. The animals, particularly the baby calves that arrive each year, need my oversight. While it's true the mother cows give them far better attention than any human can, the herd needs adequate food, fresh pasture, and an old mother cow needs an occasional assist at

The Transition

calving time. The herd is dependent on the caretaker, and in this instance, it is me.

Thus it is when I rebuild the old fence, I automatically think about my father and what he might do with the land. I understand now why he left the farm so sparingly. During the springtime calving period, his mother cows needed him. And, during the summer, a drought might set in, causing the grass to dry up and wither away, demanding a rotation of pasture for the herd. And in the cold of winter all grass disappeared and it was always necessary to feed hay. His life was tied to the cycle of the herd. Through it he gained a knowledge of life and soil, dependence on the weather, and the value of his own decisions afforded only to people who farm. Although he had plenty, he did not need all the stuff which ties me to commercial life. Later in life, he was the happiest when his grandchildren were near him. These were lessons I did not fully comprehend when he was alive. But now that he is gone and I am placed in a similar situation, I am gaining in understanding. Each old post, each strand of rusty wire is a reminder of his presence.

So it is when I rebuild that old fence along the road. The job goes best when my nephew Bob and neighbor Leo come to help and we can talk and laugh and reminisce and be neighborly while we work. We try to keep the line of posts straight, the wire drawn tight, and the strands evenly spaced. I have discovered that my father was correct about the old hedge posts we set years ago. Many are not broken off, nor are they all decayed. Many of those old hedge posts that we set so many years ago are only leaning over. They have outlived their holes. They may be weathered, but they are still useful, and so I have kept several of those old posts, interspersed along the long line of new metal T-posts. Undoubtedly, the neighbors wonder why I allow old posts in the middle of an excellent fence.

Baby Calf Time

Springtime is baby calf time in the Ozarks. The hillsides are dotted with calves, curled up, heads hanging over one shoulder, presumably fast asleep while nearby a mother cow, gently chewing her cud, stands vigilant watching over her newborn offspring.

Perhaps just a few hours earlier, the new mother wandered away from the herd to a far corner of the pasture where she labored through the tugs and pulls of birthing to deliver the new offspring. At first the calf is a slimy mess but, after mother's cleaning and cajoling in the first hour of life, the calf, teetering, labors to its feet and punches at mother's rear section until it begins to nurse—a sure sign that the newborn is healthy and will grow during the warm days of spring.

Normally, the cow/calf pair is inseparable for a few days with mother cow keeping to the fringe of the herd, munching grass and standing over her baby while the new calf spends its time sleeping, nursing, and moving in a wobbly trot, trying to keep up with its mother as she ambles from clump of grass to clump of grass.

All is not peaceful. The mother's instincts are ever protective. Woe be it, for example, to a neighbor's dog that might wander into the area. A mother cow weighs several hundred pounds, and her hoof at the end of a powerful kick is deadly.

Cows and calves seem to have been a part of the Ozarks since the first settlers arrived. I have a friend who tells about the days when open range prevailed in the Ozarks. In those days the herdsman would control his cattle with the salt lick he maintained. He kept his cattle at his homestead during the winter and then, right after the springtime calving season, the herdsman would call his cattle and lead them into the woods. There, he would place a feeder of salt near a spring for the cattle. The cattle knew to stay by the salt lick until cold weather.

Baby Calf Time

Once a week, all summer and fall, he would carry a bit of salt to the feeder and call his cows. Soon, one after the other, the cows would appear out of the Ozark forest. He would count them, see about their health, and give them another feeding of salt.

Cattle raising, of course, has changed significantly. Now it's done more scientifically. Pastures are fenced, herds are monitored for calf size at birth, rates of gain on baby calves, and even something the experts call "frame" size. Herd sires are scrutinized: long legs, straight top lines, huge hindquarters, and wide girths are desirable traits.

The Ozark landscape lends itself to beef herds. The fertile upland plain supports a high concentration of cattle; the rough Ozark hillsides are good for little else. And beef production has an exponential sequence to it: one bull, thirty cows; two bulls, sixty cows, and so on which allows beef herds to fit the small and medium-sized farms of the Ozarks. Of course, nothing on a farm fits into any formula, but it's a good guide. And, the Ozark farmers raise lots of good hay—the major roughage necessary to get beef cows through cold winter days. No wonder beef cattle flourish in the Ozarks.

But the land itself and the cattle are just the basics of beef production. A good herdsman must know about soil fertility, the condition of the land, machinery costs, necessary grain purchases, and the fluctuations in beef prices. When he has gained this essential knowledge, he knows what he will need to raise a herd of beef cattle. Some Ozarkers accomplish this and are skilled enough to make a living with their beef herds. They have applied their knowledge about the industry so that, at the end of most years, they make enough profit to support a family. I envy these people. Although I have a farm, I don't make a living off of it. In fact, truthfully, I have difficulty making a profit from it.

Return to the Farm

The fences need rebuilding, the tractor breaks down, a cow dies, and, unknowingly, I sell cattle when prices are down. My farm, like many in the Ozarks, is a not-for-profit endeavor, even if I did not plan it that way. Even the income tax deductions are not enough to justify ownership. Hopefully, eventually, though, the farm will supplement our family income.

If I were real honest, I would probably have to say that my farm is a luxury, and maybe a luxury that I should not afford. Still, the other day I walked across the pasture and found, back away from the herd, a mother cow standing vigilant over her new-born calf. The cow slowly walked away as I approached her, the tiny calf wobbling alongside. Other, older calves frolicked in the distance. Within a few minutes I heard a honking chorus from the sky as a huge v-shaped flock of geese passed overhead—apparently calling encouragement to each other as they passed. The sky beyond them was a perfect blue and beyond them I could see, literally, into eternity. Underneath my feet in the middle of the pasture, barely noticeable, was a huge bed of tiny fragile purple flowers, the leaves, petals, and stems hardly a half inch off the ground. I'm sure they would be gone in a few days and would only reappear again the following spring.

It makes no difference to me how much turmoil is in my day or what problems trouble my mind, give me an hour walking across my small farm, and I am righted. Clarity comes and I am at peace with myself and others.

Grinding Grain

Jolly Mill, a huge old three-story grist mill, still stands next to the water's edge on Capps Creek, the next creek over, about five miles south of our farm. It was an adventure, as a kid, to go to the mill with the family on a Sunday afternoon and see the old thirty by eighty feet three-story structure sitting on its limestone foundation and rising skyward at the base of the mill pond just as it had since before the Civil War. The mill was the only building in the immediate area, dominating by its very existence all that was around. The native siding was weathered and we could plainly see the hack marks of broad axes and the kerf marks of hand-pulled saws on the beams and timbers that held the building together, evidence of the tremendous effort of slaves to construct the building in the 1840s. Handmade mortised and tenoned joints held the building together. Occasionally we would see a trout break the surface in the mill pond or, very rarely, a muskrat or mink on the opposite shore. We could catch crawdads with our bare hands in the swift-running stream below the mill. My, it must have been a sight when it was constructed. On our visits we imagined ourselves in another time, another era.

But even fifty years ago the mill was showing signs of great deterioration with holes in the roof and pieces of siding dangling from the outside walls. Graffiti marked the building up as high as young hands could reach. And the whole structure, badly in need of renovation and repair, would creak and groan.

The old building survived the Civil War, although the town that surrounded the mill, a settlement of sixteen buildings by the name of Jollification, was burned by marauders of one side or the other. Local conjecture has always been that the mill survived because back in the pre-Civil War era most of the grain ground at the mill went to the distillery, which was located

Return to the Farm

inside the mill. The distillery was the main reason for the existence of the mill anyway. To brew whiskey you had to have mash, a concoction made mostly from ground corn. And grist mills grind corn, the main ingredient for mash—and thereby whiskey. In Civil War days in this community a nip of good liquor for soldiers to drink was more important than houses for people to live in. So, the settlement was destroyed, but the distillery and mill survived intact.

It took the tax collector to make the distillery go away. Visible traces of the distillery disappeared long ago when the government began collecting taxes on whiskey in the 1870s, although local tradition has it that even through World War I and maybe even longer, if one was well connected locally and very careful, he could obtain a quart of local brew from those who operated the mill.

After the Civil War, Jolly Mill became a viable mill again, but this time only for grinding grain for local farmers and the commercial flour market. When electricity came, electric powered grinders overtook the flour market in the late teens and early 1920s, and the old mill, despite the best efforts of the owners, lost its economic viability and slowly began to deteriorate.

The old mill, although largely a relic of the past when I was a kid, still worked and the owner/operator, Frank Haskins, liked to grind grain for anyone who would bring him corn or oats or barley. The great grinding burs—which ground ever so slowly and meticulously—were powered by an under-shot water wheel that ran, true to its name, in a horizontal position beneath the building rather than sitting upright beside the building. The operator would control the water wheel by opening a watergate and allowing water to pass under the floor of the building. The current of the released water forced the water wheel to turn. The grinding grits, wheels, and pulleys throughout the building would begin to turn, ever so slowly, and the

Grinding Grain

building would shimmy and shake, emitting groaning and creaking sounds of belts and pulleys as the grinding burs gained enough thrust, speed, and power to grind the grain.

We assumed at the time that the mill was about done for, that rain would eventually leak through the dilapidated roof to rot enough timbers so that the old mill would eventually collapse, or that doped-up kids on a binge would carelessly set it on fire on a Saturday night—or something would happen to it. We did not expect it to remain beside the creek much longer.

But it was a wonderful reminder of the past. When folks talked about it, they spoke in a wistful way, about how tough times must have been years ago, and how nice it would be to raise the huge sum of money that would be necessary to restore the old mill to its former condition. Folks realized that restoration was an idealized dream, not a common sense expectation. Eventually the mill, we thought, would break and crumble in a flood and wash down Capps Creek.

It was common sense, too, that guided Dad's decision to take grain to the mill at Pierce City to be ground into dairy feed. While he did take grain to Jolly Mill a time or two, it was mostly for novelty. The process was just too tedious for a farmer making a living off the land. The mill at Pierce City had the powerful electric-powered hammer mill that tore right into grain, immediately pulverizing it. And Dad could purchase the additives necessary to make a complete dairy feed. The process took time, but not nearly so long as waiting for the water-powered grinding wheel to grind grain at Jolly Mill.

Taking grain to town was not a particularly difficult job, but it was time consuming. It took Dad a half day or more to shovel the pickup full of grain, drive to town, and sometimes wait for another farmer or two to complete his grinding, before unloading, waiting for the grain to be ground, augered into

Return to the Farm

gunny sacks, then loaded again into the truck for the drive home.

I often went along with Dad to the mill, and I became acquainted with the half-dozen or so men who worked there. In the summertime it made little difference what day we went. Any day, to me, was a nice diversion from other farm work. But, during school time, we could go only on Saturdays. Dad found that he did not always have the half day every two weeks to take grain to town since the task interrupted his regular farm work.

And so, one day when he was particularly busy, he asked me if I would take the grain to town. I was delighted to do so. If memory serves me correctly, I was only twelve years old the first time I took feed to the mill by myself. Our first pickup, the Ford, had quit running and so Dad purchased another, which also was old, even when he purchased it. But the truck ran well. The bed of the pickup was made of lumber and the gear shift was anchored to the floor but extended way up almost to my eye level. I was just tall enough to reach down and touch the clutch with my left foot, jam the gearshift into whatever gear I needed, and then, because I was still so small I could not see over the tall steering wheel, carefully position myself on the seat and tilt my head so that I could look under the the top of the rim of the steering wheel but over the hood of the old truck to see the road. Then, off I would go to town. I was growing fast, though, so I would soon be able to see over the steering wheel in front of me.

Most of the trip to town was along dirt back roads so I had little fear of being picked up by a state patrolman until the last one mile into town. I always shuddered that last one mile, wondering if I would get stopped. I would drive straight to the mill, shovel the grain into a waiting hopper, wait for it to be ground, load the grain that was now in heavy gunny sacks, and

Grinding Grain

hope I didn't get caught in that one mile until I could reach a back road. Somehow, I was fortunate. I never did get caught.

Dairy cows need lots of grain and we grew most of our grain on the farm—barley, oats, and corn. What little wheat we grew, we sold as a cash crop. The rest of our grain and hay we used as dairy feed. We stored the grain in small granaries on the farm and, when we needed it, would use a mammoth shovel to scoop some of each of the corn and oats into the back of the pickup. Next Dad would give me a list of the additives for the ration, usually molasses, cottonseed, bran, and nutritive supplements to complete the feed. Finally, it was off to the mill to have the grain ground.

The old mill in Pierce City was owned and operated by a man named Stiffy. The local story had it that Stiffy garnered his name when he was but a child; a wasp stung him on the neck and it became so stiff he could not turn his neck for several days. The motion returned to his neck, but his nickname stuck. Although he was an older man, everyone, even us young kids, called him Stiffy.

John worked at the mill, too. John and his wife Ethel lived on a small farm a couple of miles from our farm and we were neighbors, saw each other at church, and sometimes visited at neighborhood functions.

John and Ethel were an older couple and had no children. Ethel, a very tiny woman who hardly ever spoke, gardened and canned food and kept an immaculate house. Ethel never learned to drive, although she lived several miles from the nearest town. She was alone all day and, on the week-ends, she and John would go shopping in the family pickup and to church. John was an "even keel" worker. He did not work fast, but he was always there, working in blue patched overalls, the patching made necessary because of the constant rubbing of feed sacks against his legs. He was the steadiest of workers, ready to help, and he offered a kind word to everyone. We always spoke of John and Ethel as gentle people.

Once at the mill my job was to unload the pickup, again with the big scoop, shoveling the grain into a shaft that led directly to the basement where the hammer mill was located. John was usually the one to flip the switch to start the grinder. I would hear the hum of the grinder as the whole building began to vibrate ever so slightly, followed by the sound of grain hitting the grinders. In a few minutes, the ground grain would start pouring down from an overhead spout and the workers of the mill would take the gunny sacks I had brought and fill them, one by one, with the freshly ground feed before loading it back onto the truck.

The men always bragged about how strong I was becoming, so I always jumped in the middle of the work and helped them drag the heavy gunny sacks of feed across the wooden floor, polished shiny slick because of the sliding of gunny sacks. Because of the added supplements we always had more feed to take home than we took to the mill. Grinding grain was a routine, repeated biweekly, throughout my childhood.

The sacks of freshly ground feed were heavy and the first few times I took the grain, when I was not yet a teenager, I would tug them across the board floors. But by the time I became a teenager, I could hoist a sack onto my shoulder, walk to the truck, and with a flip of the shoulder, throw the heavy sack exactly where I needed it in the back of the pickup.

After I left home for college, Dad changed his dairy operation, hoping to cut his work load while increasing his income. First, he purchased several additional milk cows, then began purchasing most of his feed and having it delivered to the farm, no longer needing to take grain to the mill to be ground. Many other farmers must have joined this trend for, one day, Dad told me that Stiffy had announced his mill would

Grinding Grain

close right after harvest. John and the other workers at the mill would lose their jobs. An old Pierce City landmark would be gone.

On another of my visits home, Dad said that the mill had closed, then had burned one night. He looked troubled when he told me that John had burned the mill. On the night of the fire, John's pickup had been seen there, and the police asked him if he knew anything about the burning of the mill. John just said, "I burned it." Since John and Stiffy were good friends, Stiffy did not want to file charges against John, but the Sheriff's department and the prosecuting attorney and insurance companies did. John was convicted of arson and sent to prison. We never understood why he did it.

Ethel continued to live alone on the farm. Mother would sometimes take her up to the state prison to visit John. Ethel was reluctant to ask for the ride since it took a day to get to Fulton and another to come home. John was in prison only a short time, however, before he developed cancer. He lingered only a few months, as I recall, and Ethel only saw him another time or two before he died. After that, it wasn't long before she died, too. In death, the two were united again in the neighborhood Berwick Church community cemetery.

I don't think many farmers grind grain anymore. Most purchase their feed, although they may raise their own roughage—hay or silage.

Strange, though, how things work out. The old Jolly Mill is still standing today and is in the best shape of any time, perhaps, since it was built. The family who owned it when I was a kid sold it to a developer from the West Coast. This gentleman, who had an affinity for the past, gave ownership of the mill and a "ton of money" to a local group who would oversee the restoration of the mill. Under the guidance of this local group, the mill has been refurbished, many of the old planks

Return to the Farm

and timbers and joists have been replaced, and others inspected and pronounced as good as they ever were. The mill pond has been cleaned out and the dam repaired. Other improvements are being made. An old-time village, complete with schoolhouse, play yard, and saw mill has emerged around the old mill. Trout, too, still jump in the mill pond.

Last year at Christmas time we attended a neighborhood celebration at the mill. The old structure was decorated with lights, some of the neighbors were dressed in clothes reminiscent of the turn of the twentieth century, candles burned, women served cookies and wassail, and a trio sang Christmas songs in the old mill. It was a wonderful nostalgic trip into the past.

The area is becoming ever more vibrant. The Missouri Conservation Commission has now purchased several hundred acres of land along Capps Creek near the mill and is slowly restoring it to wildlife use. A good road has been constructed from Highway 60 to Jolly Mill. Tourists, as well as local residents, are finding the community. I suspect the mill will become the centerpiece of the area, and few people probably realize what changes are in store for it. Lots of people now realize the value of looking back at, if not returning to, the past, and Jolly Mill helps people reminisce. Perhaps Jolly Mill will again become what it was before the Civil War, the center of community life.

As for me, my beef cows have little need for grain. In the coldest of winter I feed them some high protein feed mixture, but during the rest of the year they must forage for grass or I give them hay. Thus, I have never had any grain ground, nor probably ever will. When I do need feed, I go to a feed store in Monett in my little foreign-made station wagon and order a few sacks, a half-dozen at most, and one of the workers helps me load them. I laugh and tell him I am one of the

Grinding Grain

few farmers who can farm out of the back of his car. I suspect they are still chuckling after I leave.

Mother

I suspect that in many Ozark communities, families still find religion and neighborliness so indistinguishable that the same kindness and community traditions that bring families together in life also sustain them in times of sorrow and death. This story is of an oft-repeated but little-told Ozark scene. It is the story of the death of my mother Pauline McGill, the Berwick community, and our family home which stands within sight of the Berwick church.

Several Januaries ago, we drove the winding roads across the Ozark hills and through the valleys to my mother's eightieth birthday party. It was Saturday, the day her children and grandchildren had chosen to celebrate the birth of this resolute, God-fearing mother and loving grandmother. On the way home, my wife commented that Mother, despite the fact she was legally blind, appeared to be in excellent health and might live to be ninety or beyond.

Sunday, members of her church would honor her again—this time with a dinner, card shower, and community sing. She stood at that service and testified about her Lord and gave thanks for her many friends and neighbors in the Berwick community.

The following day, Monday, on her actual birth date, my mother doubled over with an excruciating pain that originated from her pancreas; a grandchild rushed her to the hospital in Monett. After a quick examination, the doctor immediately had her transferred to Cox South Hospital in Springfield. But all the tests, machines, tubes, wires, nursing care, medical specialists, surgeons, operations, and intensive care units could do little to influence the outcome of her life. Within a week she succumbed, and life passed from the resolute little lady.

Before she died residents of the Berwick community, led by the pastor of her church where she had been a member

Mother

for so much of her life, visited the hospital, wished family members well, and reminisced about neighborhood picnics and church gatherings. They told us how much she gave to the community even though the neighborhood had changed drastically in recent years as long-time residents left and new neighbors moved in. With her death the outpouring of love and honor mushroomed. A steady stream of neighbors visited the home and brought food for the family. The conversations with these neighbors centered around the active and varied nature of her life. Some told about her activities in literary and civic clubs, where she had been a long-time member. Others remembered her helping serve dinners at neighborhood farm sales. One woman, who still makes all of her own clothing, stated that my mother years ago taught her to sew. But all conversations finally connected back to the fact that her church was central to her life.

Visiting hour was at the funeral home, on Friday night, the night before the funeral. Many friends came to express sympathy and tell of their own bereavement at the loss of a loved one. Most had a personal understanding of loss—they too had experienced the death of family members and friends. One friend opined that when the last parent dies, the children become orphans, and to become an orphan, even at 50, means that new family supports, customs, and traditions must be created. The cornerstone of the family foundation was gone.

One change from the past had to do with the grave site. When I was a boy, the men in the community gathered to dig the grave and talk about the deceased. Now, the grave was "opened" by an operator on a piece of high-powered machinery. But the result was the same. Mother would again lie next to my father.

Saturday, the day of mother's funeral, an unrelenting cold wind blew from the north, lowering the temperature into

Return to the Farm

the twenties. Family members from other states arrived in the morning until our family farmhouse bulged with children, grandchildren, cousins, nephews, and nieces. Late in the morning we formed an entourage and drove to the church where friends from the community had prepared a feast in the basement of the church. Her family and close friends ate together, talked, and laughed in nervous anticipation of the service which would soon be held upstairs.

Finally, it was time to file upstairs, past a standing congregation, to pews directly in front of the pulpit at the front of the church. The church overflowed with people. Flowers filled the front of the church. A woman at the piano began to play and sing "Oh God, Our God." Other family hymns, "Amazing Grace" and "It is Well" were sung. The hymns were the same ones played at my father's funeral.

The minister preached, not a modern sermon about the rewards of doing well, but a "scriptural" message straight from the Bible. The man of God spoke plainly. He reminded us that Mother was already in heaven rejoicing with my father and other neighbors and friends who had gone on before. It was plainly written in the Word, he reminded us, that her heavenly Father had taken her from Berwick to another, even better community of believers. He renewed our hope.

Thirty years ago Mother had taught a Sunday school class to teenage boys in the Berwick church. After the funeral service, members of that class, now grown men with adult children of their own, served as pallbearers and carried her from the sanctuary to the cemetery beside the church. They passed beneath the mammoth old white oak tree at the entrance of the cemetery, around stones where the loved ones of many mourners lay, and to the open grave in the family plot where she was laid to rest. What a tribute to her life—to have those to whom she taught her faith carry her to her grave.

Mother

Years have passed now since these events took place. We still miss Mother.

Charley and Me

"Get bigger calves to the table faster and you'll make more money," the experts have told farmers for the past decade or so. And the experts have developed plans to do this. "Crossbreed your cows and they will grow faster with the improved genetics. That's how you'll make money," they say. And the experts have been correct—about the vigorous growth, at least. So, instead of keeping purebred beef herds, more farmers have crossed exotic bulls, some of them imported, with the popular strains of cattle of bygone years until now cow herds of various sizes, colors, conformity, and even ear length dot the countryside. (Nowadays, we want calves with long meaty legs, a reversal of just a few years ago when short-legged calves would supposedly produce the best carcasses.) This interest in genetics has developed bigger calves that get to the dinner table faster—but the farmers are rarely making more money.

But while the improvement has taken place, something else has happened, too. Today's beef cattle have become not only more vigorous in growth but also more vigorous in demeanor. True, over the years there has always been an ornery cow or two in each herd, or the herd bull has been of a cantankerous sort, or a particular young calf had a "wild" streak in it. Now, though, thanks to all the crossbreeding, many of today's bigger cattle are nervous and sometimes uncontrollable. Some of them, in fact, are just plain mean-spirited, wild, so that it is no easy task to corral cattle to feed, medicate, or sell.

When I was a kid, we were personally acquainted with each cow. When it was necessary to sell an animal, Dad and I could drive the animal to a small holding pen located next to the road at the corner of our barn lot. Dad would back the old Ford pickup down the road to the embankment at the holding

Charley and Me

pen and pull open the wooden gate separating the pen and pickup. Then, sometimes the whole family would gather around to say good-bye to the animal before Dad would give the cow one last pat on the rump, comment on her contribution to the farm, loop her tail over her back and give it a sharp twist. Normally, the unsuspecting animal would jump into the pickup, ready to make her way to market.

But with the temperamental cattle of today, "working cattle" has changed. I discovered I had developed a lingering fear in the back of my mind about getting run over by one of these wild 600-pound pasture-fed calves that had never been inside a holding pen before, or a 900-pound cow anxious about being separated from its baby, or a one-ton-bull. And, while most are slow to admit it, I discovered other stockmen, some of them long-time cattlemen, with the same concern. Caring for a herd of cattle, particularly with makeshift pens and corrals, is dangerous. Thus it was that I came to want a good corral.

The sturdiest corrals, I knew, were made not from native posts and lumber, but from heavy pipe which, around here, is usually obtained as used pipe from the oil fields of Oklahoma. A touch of pride welled up in me. I decided I needed the very best corral possible, one made of heavy pipe. After all, it would be nice to have the "very best" of something.

I ordered a mail-order booklet of plans from Iowa and while I waited to accumulate the money to build the corral, I studied the plans—deciding how best to modify the corral for my farmstead. Soon paper clips marked the pages I found most pertinent. The plans indicated the corral should be a rectangle approximately 60 x 120 feet consisting of three holding pens, a major runway leading to the pens, and lots of gates for sorting cattle. In addition, one end of the runway would be circular and made into a working chute with added pipe for strength and a massive, self-catching headgate to secure the

Return to the Farm

animals. It would take well over a thousand feet of used pipe and two dozen or so heavy gauged woven-wire panels to build the corral. The completed corral would be easy to use and much more than adequate for my small herd. Above all, it would allay my fears of being run over by an excited animal.

Not having money budgeted for the project, I wanted to pay for the corral out of the semi-yearly sale of calves. One year passed, then another and I still had no definite date set for building the corral. Then, at our annual Christmas party last year a new friend, Charley, from Wichita, Kansas, who is a welder by trade, surprised me by offering to come and weld the corral for me. "It will be quite a task," Charley said. I readily and gratefully accepted his offer, even then greatly underestimating the time and effort he would give to the project. We set early April as the time for his arrival.

For several months I studied the plans even more, spent time at junk yards collecting used pipe for the endeavor, and hauling the heavy loads to the farm. My old GMC truck, with holes in the muffler and the engine clattering, seemed to collapse as I drove it in the yard from the last trip. It, too, was saying, "begin."

The plans called for sixty-five pipe posts, at a height of six feet, to be set into the ground. The corral would have to be laid out in the pasture, holes marked, then dug, pipes set in the poles, tamped in, and concrete added around the poles. I told Charley I would have all sixty-five posts set before he arrived.

A month before Charley was to arrive, I purchased a mechanical posthole digger to mount behind my tractor. It would be invaluable in digging those 65 holes but, lo and behold, the hydraulic lift on my tractor failed and would not pull the digger from a dug post hole. The digger just sat in the first completed hole I drilled, still turning, but I could not get the tractor to lift it out. The hydraulic system on the tractor

Charley and Me

was broken—kaput—and would not lift. Frustrated, I called a tractor mechanic. He vowed over the phone he knew the cause of the problem. He could replace two valves in the hydraulic system, which he knew would be defective, and in a couple of days and for $120 he would have it working again. "You'll be back to work in no time at all," he declared.

It was almost three weeks later that the mechanic called and reported that the tractor was repaired and the bill would be around $700. Counting the costs of the mechanical digger, I had over $1,000 in that first hole. I was also three weeks behind schedule before I started. Nor were these repair costs in my stingy budget.

In the middle of April, Charley arrived in the yard driving a late-model pickup truck and pulling a moderate sized trailer which he had constructed. The trailer contained a neat array of toolboxes, a generator, welder, cutting torch and supplies, and other needed equipment. He had a lot of tools and equipment compacted into a small space.

I hollered a greeting at him, asking if he had had a good day. "I don't tolerate anything but good days," he bellered back. I would soon learn he didn't tolerate shoddy work either, not from himself or anyone else. Charley demanded first-rate work on his job sites.

"Then let's eat," I responded.

"Don't have time to eat," he groused, not moving from his pickup. "Let's get to work. It's noon and we're a half-day behind schedule already," he continued. At that time, he was still unaware that I only had sixteen of the sixty-five posts set.

We both went without lunch that day while we worked until dark. Slowly over the next few days Charley would take tools, as needed, from the welding trailer he had pulled to the site. Charley brought out clamps, markers, a metal square, a striker, levels, power saws, goggles, large and small striking hammers, and several tape measures. His blue jacket, burned by many hot sparks, was ragged. And, of course, on the trailer, was a gasoline generator/welder which would power the construction of the corral. Many of the tools reminded me of those of a blacksmith I had once watched—whose job also was to fuse metals together. Except Charley's tools were larger and, instead of a forge and anvil, he had a modern-day welder and cutting torch.

Charley knew the procedure for building pipe fences and corrals. He had done it many times. Although I thought I knew something about building barbed-wire fences, I had never built a fence with pipe. I was a complete novice. My inexperience added to Charley's workload. Not only did Charley help me set the pipe posts for the perimeter of the corral, he had to teach me how to set the posts, firmly anchored in concrete, in a straight line. The steel pipe was heavy, the work exacting. The weather that first day and for the next two weeks was cool, perfect for heavy outdoor work. We accomplished much that first afternoon, as we would every day that we worked.

Charley taught me about the processes of working with pipe as well as the "tricks of the trade" he had learned in the past forty years. He even had "tricks" for what I thought would be the simplest of tasks.

When laying out one side of the corral, he showed me how to pull the twine taut and then tie it in a simple bricklayer's knot so the line would stay tight, but the knot would come loose with a simple tug of the twine. "It's easier to keep the posts aligned when you do it this way," he would tell me. When we were digging postholes, the clutch on the tractor malfunctioned. Immediately, Charley knew the adjustment to correct the problem. He then showed me how to set the heavy

Charley and Me

steel posts into the hole with a minimum of effort, to "bump" the posts to get them perpendicular in the hole, then an easy way to open the heavy sack of concrete so the gray granules would pour freely into a posthole, to tamp the air out of the dry concrete mixture before adding water, and how to pour water into the hole to harden the concrete that would hold the posts for another twenty-five years or so.

Charley also wasted no time in teaching me to use a cutting torch. Since each post varied in height, each post had to be cut to the same height to make the neat-looking fence that was the hallmark of Charley's work. He handed me the cutting end of an acetylene cutting torch, then a striker and the control knobs of the torch and showed me how to ignite the striker. In just a few minutes, he had me cutting the pipe at exactly six feet above ground level. On the very first day I learned to cut pipe with an acetylene torch; before the week was out, I would learn the basics of welding.

Usually, I would work a few paces in front of Charley, marking the perimeter, digging the holes, and laying out pipe, and, with Charley's suggestions, setting the posts according to the plans I had. Behind me, Charley would be straightening pipe, hoisting the heavy pipe for the top rail, and tacking it in place to be welded later. Even the slightest deviation from dead-straight would bother Charley, I discovered. "Do it right, now," he would say many times, "and it will last longer than you or I will." He knew exactly what I was striving for. I did not want to rebuild this corral.

There were some problems. When we came to the one rounded corner of the corral that would serve as the working chute, I could not get the dimensions of the corral to fit the dimensions of the plans. "There's a mistake in the plans," I yelled to Charley. "What kind of an engineer do you suppose drew these up?"

Return to the Farm

"You can tell an engineer, but you can't tell him much," Charley yelled back.

Charley proceeded to tell me several stories about other error-ridden plans he had encountered over the years and the efforts it took to correct them. We fixed the rounded corner Charley's way, and it worked.

Because of the extra bracing necessary to make the rounded working chute, progress slowed as we came to that corner of the corral. Again, the ambivalent plans were unclear on how to proceed with the bracing. Charley was perplexed one evening as we talked about how to brace the rounded corner, a task we would need to complete the following day. Early the next morning Charley was all smiles. "I woke up in the middle of the night and had figured out how to do that bracing while I slept," he exclaimed with a smile. Then he drew a diagram of the bracing. Clearly his idea would work.

Perhaps, though, the best part of building the corral was traveling to and from work each day. Our residence is an hour's drive from the farm and so, after work each night, Charley would drive the two of us to and from the house in his pickup truck. It was during these hours of riding and talking that I learned more about Charley. I had already met his wife, Jane, and two children and grandchildren. Charley confirmed what I had assumed, that he had worked hard all his life, had only had two employers in that lifetime of work, and his life work had consisted of driving heavy machines and welding. He was also devout in his religious beliefs, and he and Jane spent more than just Sunday mornings in church. Their faith, and helping others, gave them a fresh outlook on life. I discovered, too, that he would drink limeade and so a ritual began. Every evening when we passed through Monett, we would stop at the local fast-food restaurant and order the biggest limeades we could get. The young attendants would squeeze

Charley and Me

the juice fresh from limes. My, at the end of a hard day's work they were good.

    Our daily routine was simple and one shared with workers everywhere. Every morning we would arise before daybreak, breakfast on hot coffee, toast, and oatmeal and be on our way for the hour's drive to the farm. I took sandwiches and at lunch time insisted to a reluctant Charley that we eat them. This lean man was not used to eating lunch. Then, about dark, we would begin the trip back to the house. We worked hard.

    I had thought we could complete the corral with a week of hard work. But it took three days just to finish setting the sixty-five pipe posts. After that, we had to place the top rail on the posts, align them in a straight line, and weld them. Charley had done this on many other fences and corrals—and he knew it took time to do well. After a week of work, we had the perimeter completed, the heavy woven-wire panels fixed in place, and were beginning to weld the cross fencing inside the corral. Still, there were many gates to make and it was becoming apparent that we would not complete the corral during his visit. We worked ten straight days, leaving the house before dawn's light and working until dark. I had never seen a man over seventy years old work so hard. He could outwork me. But, toward the end of the week, even Charley's endurance was beginning to wane. He had already stayed longer than he had thought would be necessary. He talked about going home.

    We worked until almost total darkness Thursday night. Charley pushed himself to complete a couple of gates necessary to make the corral operable. I told Charley I could use make-shift gates and get the cattle in during the summertime. Charley promised he would come back in the fall and complete work on the gates. We spent Friday morning cleaning up the premises, stacking short pieces of leftover pipe, and completing a final weld. All the tools were placed neatly inside the

Return to the Farm

toolboxes, and the cutting torch was fastened securely to the trailer. The welder showed no signs of use. When Charley was ready to leave, the trailer looked exactly as it had when he arrived.

Charley would only take payment for the gasoline he had used in his pickup and for the welding rod burnt in the construction of the corral. He would take no other payment. He said he had lived long enough to know that money is not a primary interest in his life now. Still, I know he and Jane are far from rich. He's just a different kind of person. I am glad to have him for a friend.

Now, out in the middle of my pasture I have a nearly completed corral. I've worked on it some since Charley left and it serves its purpose well. In the fall, Charley will return and we will complete the gates. But already, every time I see it — this will continue for the rest of my life — I will think of Charley.

Note
Charley did return last fall and, except for minor adjustments, we have now completed the corral.

Doc Charley
and Other Doctors I have known

My, how medical care has changed.

When I was a kid, we used to go to Doc Charley Spear, whose office was on the north side of the street next to the bandstand and across from the bank in Pierce City.

Back then, when one of us kids got sick, Mother would, without an appointment, take us to Doc Charley where we'd take a seat on one of his dilapidated chairs or hard benches and wait. Soon, either the nurse, who was his wife, or the doctor himself would come out and invite us into one of the back rooms where he would calm our fears, look at our affliction, and send us on our way—usually with medicine he had both prescribed and dispensed. The adults in the neighborhood, my parents included, used to laugh and say that he was so tight he couldn't give a prescription to be filled by a pharmacist. He had to fill them himself because he had to make all that additional money. We'd give him the three or four extra dollars for the prescription and be on our way. Anyway, we liked Doc Charley and used him for our family physician.

My major childhood calamity came one hot summer day when one of our cows, a balky one, had gotten out of the pasture and did not want to go back in. I was a determined teenager wanting to prove to that cow that she should do as I wanted, and since I believed myself to be a nimble young man with athletic prowess, I began chasing the cow. We were both racing at full speed. Just when I was about to overtake the cow in a corner of the barn lot, she whirled and both of us got caught in the fence. She escaped, but I went sailing head over heels and landed on my arm—fracturing it at the ball where the arm connects to the shoulder. The pain was excruciating. The neighbors arrived shortly thereafter and I was loaded into the

Return to the Farm

back of their car and taken to Doc Charley, who x-rayed the break and then could only shake his head. Again, I was loaded into the backseat of the car, still in pain, and we headed to a hospital in Springfield where they could "fix me right." Besides, I was told, the same doctor who would work on me had worked on Mickey Mantle.

When I awoke from surgery the following day, I had a body cast that extended from waist to neck with my right arm extended outward as if I was preparing to flap a wing and fly—except it was held rigidly in place by a brace that went from my wrist to the bottom of the cast, about belly-button high. I wasn't going anywhere.

All the next morning the nurses cooed over me and stood in line to sign my cast. "Wasn't I a brave fellow?" they exclaimed. My parents arrived about noon, loaded me into the family car, and took me home to recuperate. This, of course, effectively stopped the attention I was getting from the nurses.

In high school in the fall I again was often the center of attention, this time from my classmates. The girls wanted to sign the bulky body cast. I had the utmost respect from the guys, too. With such a large cast, everyone got out of my way when I walked down the hallway. That cast would only hurt someone else if we had a collision. Besides, hadn't the doctor who worked on Mickey Mantle worked on me, too? For several weeks I was the center of attention, even if the cast was hot and restricting. Then, just before basketball season, the Mickey Mantle doctor took the cast off, which really, was all that was important. I wanted to play basketball.

Other than that incident, having some sneezes, and a cracked rib or two, not much happened to me medically as a young man. I remember going to Doc Charley another time or two during that period and until someone told us that Doc Charley had retired and moved away to a farm in Illinois.

Doc Charley and Other Doctors I have known

Over the next several years other doctors served my medical needs.

When we moved to our present location from St. Louis over 20 years ago, I had to select a new doctor as my personal physician. "Get a young doctor," I was told by someone, "so he'll still be around when you're old, and he'll know everything about you. You won't have to go around switching doctors all the time." I tried to do this. Somehow, I came upon a young doctor who was enthusiastic and energetic and undoubtedly much better schooled than Doc Charley. He was with one of the most prestigious "medical groups" in Springfield at the time. Since he took the medical insurance of the company where I was employed, he met my requirements.

This new doctor is still the one I go to, and he has looked after my aches and pains, both real and imaginary, over the years. Except for occasional bouts with depression, which have about been eliminated now, I've stayed "healthy as a bull" over these years. I think he's done an extremely good job of keeping me together.

However, a visit to this doctor, even the first one, was much different than those to Doc Charley many years before.

For starters, I began getting appointment letters telling me the date and time of my yearly appointments, not to eat or drink after midnight the night before going in to the physician, and that I should arrive early in the morning to have some blood drawn. When I arrived in the morning, a medical assistant would stick dabs of gooey paste on selected parts of my chest, hook some wires up to me, tell me to run in place, and monitor my heart. Finally, when I returned later in the afternoon, I would see the doctor who would look at the test results, talk to me about my health, and examine my body parts.

This new doctor has always greeted me with an affectionate smile and handshake. His examinations have always

Return to the Farm

been thorough. Normally, the care he has given me has consisted of this yearly physical.

But, the health system has changed. I should have know that health services would dramatically change when, twenty years or so ago, I arrived for my yearly checkup and had trouble finding his name on the office door. The office administrator, CPA trained no doubt, had her name plastered in huge letters across the top of the doorway. The names of the partners, MD's all, were printed in small letters at the bottom of the door. This was the first time the receptionist requested my insurance card before she asked me my name.

The next time I went to see this doctor, his practice was in another, larger suite in a much larger medical complex. The clinic had grown in the number of physicians, too, or, at least, the photographs of more MD's hung on the wall, although he was normally the only physician I saw in the back corridor where his examination and consultation offices are located. Change in his medical practice never ceases. Recently his location again changed to a still larger suite nearer the hospital. Over the years he's added a personal nurse to help with patients and not too long ago I met a physican's assistant who can also screen patients and diagnose minor problems that might arise unexpectedly.

The clinic seems to teem with medical personnel, with doctors, nurses, and lab technicians, all wearing medical uniforms of some kind, hurrying to and fro. Maybe they take their cue from a bee hive, but, I suppose, everyone back there knows what they are up to at any particular time. It just looks so antiseptically clean and orderly—not at all like Doc Charley's.

Even the waiting room is so much different. Doc Charley's office only had broken down couches and hard-back benches. The pictures on the wall were not even straight, and often there was no receptionist to greet us. Everything in this

Doc Charley and Other Doctors I have known

new waiting room is shiny and new and comfortable with plenty of magazines lying around.

But those seats in this new doctor's office are not easy to get to. The first time I went to meet the doctor in this, his latest office, the neat and attractive receptionist sitting behind a huge counter met me with a cold stare as I stepped off the elevator. "Show me your card," she ordered, followed by a "your co-pay is $10.00." Once she had seen my insurance card and could call me by name, she was able to request additional information, "Who is your doctor?" she asked, followed by a, "have a seat over there" statement.

No, that definitely is not like Doc Charley's old office.

However, this doctor saw me through my most traumatic adult medical moments. A few years ago when I was building fence on the farm and needed to lift some particularly heavy corner posts, I felt what seemed to be a small tear inside my groin. The pain was grating and over the next few days I felt weak, a bit nauseous, and lifeless. The dull pain persisted.

"Oh, do I have a hernia?" I wondered.

My new doctor, although by now I had gone to him for years, confirmed my fears.

"Yep, you've got a hernia," he stated flatly, as he examined the rip in my inner muscles. "And, here's the surgeon who specializes in fixing hernias. In fact, he had one of his own, not too long ago. He knows all about them, and how to fix them. And, your insurance will cover the costs when this doctor fixes it."

Within a few minutes, I had an appointment with this hernia repairman.

And, sure enough, in a few days I was lying flat on my back on a hospital surgery table staring at this surgeon and other medical technicians, who, in turn, were obviously waiting for me to respond to the anesthesia being administered so that I would go to sleep and they could go to work. They didn't have long to wait. I was asleep almost instantly.

My major recollection of the whole surgery process was that, upon awaking from the surgery, a nurse immediately asked me if I could "go pee."

I answered "of course," tottered from the bed and, holding my side with a pillow, waddled the few steps to the bathroom where, unaided, I relieved myself.

When I emerged from the tiny bathroom, she demanded, "Did you go?"

"Yes," I muttered as I stumbled my way back to the bed.

"Aye," she said, "If you can pee, you can go home," she concluded as she somehow adroitly managed to block my progress to the bed, hand me my clothes, and point to the door.

Before I could get my street clothes on, someone else had stripped and made the bed.

I was halfway down the hall, headed for home, before my wife caught up with me. Meanwhile, I could hear another patient on a bed being rolled down the hallway toward the room I had just vacated.

This was so much different than my broken arm surgery thirty years earlier.

I still refer to this physician as "my doctor." He's matured over the years, even as his hair has begun to turn slightly gray. He seems to have learned from experience what to look for in an examination. His judgment seems very good. Still, when I arrive for my yearly checkup, I'm greeted with the same friendly smile, the affable handshake, and the perennial question, "How have you been this year?" I make the same statement about the joy of having a prostate examination. He responds with the same corny joke—something about a corn cob. We'll often spend more time talking about farming or hunting and fishing and vacations than we do about health.

But, when we do talk about health, he will often turn to the newest aid in medical help in his office, his computer. Through it, he can recite the chronological events of my last several visits, bringing up concerns I had forgotten.

The doctor now has my complete medical history in his computer. During my first visit after he acquired it, he seemed to spend more time in front of the computer trying to decide which key to punch rather than talking to me. He looked a little hemmed in, learning to use his computer while talking to his patients.

Now, though, he knows how to make that computer do what he wants, and I am again the center of attention, even if the computer is omnipresent. At times, though, he looks a little forlornly into the computer and occasionally spends more time looking into the computer than at me. It's probably more of a blessing than a curse that he has that computer. However, I'm aware of the abuses insurance companies and lawyers can make of so much readily available medical information if, in the name of profits, they find it to their benefit to abuse the medical system. But you can tell, the computer has changed his medical practice. He has numbers for every disease.

The other big change I find in my doctor's office is in the photographs on the wall. When leaving his office and waiting at the front desk for my next appointment card, I have time to look at the pictures on the wall. I always look for his picture. His picture hangs in the row with the larger frames, with the senior staff members of the practice. He ceased to be the new kid on the block a long time ago. Another row of pictures, below his, contains portraits of the newer members of the practice. These are photographs of young people, both male and female, who are the newest MD's on the staff. On the other

Return to the Farm

side of the room are the photos of the doctors who years ago were proudly displayed in the front of the office, the founders of the practice, I think. Those photos are now on the emeritus side of the room.

Over the years this doctor has helped keep me in good health, and I am thankful for his help. I did select well. But now, I'm not so sure he will be my last physician. As age creeps up on both of us, he may well retire before I expire. We'll just have to wait and see which occurs first.

But there is a footnote to this story, and perhaps the footnote is longer than the story.

One day on returning from the farm my right groin was aching. I imagined my muscle having disintegrated into ground beef. The dull pain was strong and when I touched a certain place on the groin, the pain intensified. "That feels just like my original hernia," I thought. I was busy. "I can get through this for a few days until I have time to go to a doctor," I told myself.

By the end of the week I stopped by the doctor's office while I was in Springfield on other business, not bothering with the formality of calling ahead for an appointment.

When I entered the waiting room late in the afternoon, the receptionist was standing at the far wall, chatting with another clerk. When she spied me standing in front of her counter, she approached me and asked what she might do to help. I explained my situation. "Could I see the doctor?" I asked.

She had me take a seat in the waiting room, then disappeared behind the closed doors that I knew led to the hallway to his office. Then, she reappeared in a few moments. "The doctor is still seeing appointments," she stated. "You'll have to go to an emergency room for help." She stood directly in front of me, crossed her arms, and said no more. It was plain to see that my visit was ended.

Doc Charley and other Doctors I have known

I was perplexed. Here I was, standing not thirty feet from the physician who had been treating me for years, but we were separated by a wall I could not get past. I couldn't help but think of Dr. Charley Spears.

I had a rush of work at my employment and so endured the pain—some days more and others less—for a couple of weeks until I could finally relax. I again called the physician's office and asked to see the doctor. Again, I was given the same response. "If that pain is really great, go to an emergency room," the receptionist at the other end said. "It will be a month before you can see your doctor."

Since it is a few miles closer to go to the emergency room in Branson, rather than Springfield, I ended up in the Branson emergency room on a Friday early in November where, in a curtained examination room, both a doctor and his assistant took turns pushing and probing my groin. After the examination, the two surprised me by announcing that I had not one but two hernias, one on each side of my groin, and that I would need surgery as promptly as possible. "We'll get you over to a surgeon and you can get them taken care of in a few days," they agreed. Then they shook my hand, made an appointment for me three days hence with a local surgeon, and I was on my way out the door. My groin hurt where they had been probing. But the two had been efficient and medical progress was being made, I believed. My next stop would be at a surgeon's office.

By the time I got home, there was a message on my answering machine. The surgeon's receptionist had called. She was candid. She said that without a referral, the surgeon couldn't get paid for my upcoming visit. She gave me the fax number of the surgeon's office. "Can you have your regular physician give us a referral?" she asked.

Can it be that the physician I couldn't see has to refer me to a surgeon he doesn't even know, I wondered. I called my

Return to the Farm

physician's office, was promptly connected to an answering machine to whom I explained my need for a referral, and gave it the fax number of the surgeon I was about to see.

Two days later when I arrived at the surgeon's office, both a receptionist and the office nurse greeted me. "Hello! Come in," they smiled. They copied my insurance card and gave me a long health questionnaire to fill out. "And, by the way," the receptionist said, "we need a referral from your doctor."

"But I called and asked them to send one, just like you asked," I responded.

"We don't have it", she replied, "and the doctor doesn't get paid if we don't get one. Maybe you can go to his office and get one," she suggested.

"But that's an eighty-mile round trip," I protested. "Maybe I can try to call him on the phone again," I answered.

"Good," the receptionist said, "they can usually get a referral down here in a couple of minutes after calling." By now we were talking insuranceese.

I called the phone number of my primary physician, told the receptionist my need, and she promptly put me through to an answering machine—presumably the same one I had talked to three days earlier. Again, I explained my need to the answering machine. "That should do it," the receptionist in front of me said. But, thirty minutes elapsed and still there was no response.

"You may have to go get it yet," she said.

"I'll try again," I said. I was still determined, but I recognized desperation in my own voice. When I reached the receptionist this time, I was brief. "Let me speak to a human...," I began to plead. But to no avail! At the flip of a finger she had me back in contact with the same answering machine. Again I recited my needs to the answering machine. Another thirty minutes elapsed. The fax did not ring. By now it was past noon.

Doc Charley and Other Doctors I have known

"I'll try this time," the receptionist said. Both the nurse—I now surmised that the receptionist, new to the job, was being trained by the nurse—and I listened as the receptionist in front of me explained to the receptionist at the other end of the phone line about my need for a referral—evidently a very common request in medical offices. At no time did she appear to have to talk to a machine. She did look quizzical when she got off the phone, turned to the nurse, and said, "She was rude, but I think we'll get it."

"They're usually rude at medical centers in Springfield," the nurse replied, "but you'll get the referral now."

Within two minutes the fax rang and within five minutes I was back in the surgeon's office where he poked around on my groin trying to find hernias to operate on.

"I find one, not two," he concluded. "You've got one over there on your left side, that pain on the right is something else. To ease that pain we'll just shoot your right side full of medication when you're under anesthesia. We can do it at the same time we operate. You'll be all right," he went on. "The needle we use for that kind of a shot is rather large. We'll just do it while you're under.

"Get an appointment on the way out to get your blood work done at the hospital on Monday, then we operate on Wednesday," he said as he finished the exam. "And, don't forget to bring someone with you to the surgery to drive you home afterwards." He added as an afterthought, "You'll need a driver."

Monday at the hospital, an efficient nurse who obviously had drawn thousands of vials of blood, took one from me. "Surgery is Wednesday," she reminded me, "have a driver come with you to take you home. Good luck."

Return to the Farm

But surgery was not Wednesday. Tuesday, I got a phone call at home from, if I remember correctly, the receptionist at the surgeon's office. "The blood work from the lab showed your platelet count to be way down, less than 10% of normal," she said. "You'll have to go back to your primary physician in Springfield. We already have an appointment made for you for Friday at his office for more blood work." I did not understand the potential severity of the results.

"A screw-up at the lab! Some kind of good medical work that is," I smirked to myself. "And, I have to go back to Springfield."

But, it wasn't a screw-up at the lab, and it turned out to be excellent medical work.

The technician who took the second vial of blood at my regular physician's office on Friday indicated it would be a full week, the following Thursday before I received the results back. But it wasn't Thursday, a full week hence that I received the results. The following day, Saturday, I received a phone call from an on-call physician at the medical center telling me my platelet count was dangerously low, that I should not cut myself or receive a blow to the head, that I should come to the hospital as soon as reasonably possible.

The following day, Sunday, I was lying in a bed in a private room at a Springfield hospital, feeling fine, but with doctors and nurses and lab technicians filtering in and out of my room like bees at a bee hive. For two days they punched, pushed, poked needles in me, examined my skin and scanned my torso before deciding that I had an uncommon blood disease, an immune disorder of unknown origin called ITP. My blood would not clot with such a low platelet level, they told me, and spontaneous bleeding was a possibility. However, before I left the hospital, I was given medication that it was hoped would raise my platelet level. Relief appeared to be in

Doc Charley and Other Doctors I have known

the offing. I would be under the care of still another doctor, a hematologist, for some time to come, and would have to make frequent visits to her office. "Often removing the spleen by surgery corrects this malady," she told me, while indicating that medications might cure the problem, too. She seemed to lean toward surgery, but wanted to wait a few days before making a final decision.

I tried to explain the situation to my neighbor, Paul, the week after returning home from the hospital. "I'll be fine, we think," I explained. And I recounted for him, hopefully without being too boring, step-by-step the process that had brought me to the hospital and back home again. "The doctor who didn't see me referred me to the surgeon who wouldn't operate on me," I explained, still miffed by the first lack of service from my doctor.

Paul, who himself was recuperating from serious medical treatments, just laughed. "You have a lot to learn, Bob," he said. "Here, have a beer." He smiled a reassuring smile and took over the lead in our conversation. "Let me tell you how doctors bill their patients, " he said, as he returned to his comfortable chair and explained to me the similarities in the billing practices of physicians and lawyers. "But you know what," he ended, "the medical system has always worked out for us in the end."

Intuitively, I felt he was right. A few days earlier I had arrived at an emergency room, miffed because of the extra effort needed on my part to visit with a physician. Yet, for all the gaffs and glitches and lack of timely referrals, the medical system had caught what could have led to a sudden and serious medical crisis. As it was, a diagnosis had been made and treatment started. Further, I had confidence in the doctors, nurses, and medical technicians responding to my immune breakdown. Down deep I knew that, somehow, either because of the system,

Return to the Farm

or in spite of it, that visit to the hospital added immeasurably to my length of stay on planet earth. Paul was right, the medical system had worked.

Ironically, a few days later, when I entered the office of the hematologist for the first time, I ran up against another glitch in the medical system. I was seated in a waiting room, trying to concentrate on reading a magazine but wondering, instead, if by now my platelet count was responding to medications. A very pleasant nurse appeared before me and explained that the doctor needed a platelet count of my blood before she could see me. "I'm sorry we didn't tell you to get your blood drawn a couple of days ago," the receptionist told me. But she was candid. "The doctor won't get paid if you don't have your blood lab work results here before she sees you. We can do the lab work here in the office, but your insurance company won't pay for the lab work if it's done here," she continued. "Just go to the lab down the street when you leave here, then come back in a couple of days," she stated. "Here's your new appointment card."

I would have to wait to see the specialist. "Could I have been but thirty feet from the hematologist's office?" I thought as I left her office.

I stopped off at a bar on the way home and purchased my own beer. No need to wait for Paul on this one.

The routine of frequent visits that would last for months began a few days later when I arrived for my next appointment. The receptionist confirmed that all blood tests were completed and offered me a seat in the waiting room. This time, I had time to look around. Other patients partially filled the room. Over yonder an elderly gentleman chatted with his wife. Next to me a carefully dressed middle-aged woman sitting in a wheelchair visited pleasantly with a friend. Still more patients entered, sat, waited, and left for their appointments when their names were called. Patients, some of whom undoubtedly had life-threatening

Doc Charley and other Doctors I have known

diseases much more serious than my own, visited, talked to each other, supported each other. My fears were quieted just sitting there. I found strength just listening to and visiting with other patients. The room seemed to transform itself into a serene, almost a cheery place. A literature rack filled with informational pamphlets and booklets about how best to cope with and overcome breast, bladder, and prostate cancer offered even more hope. Then, finally, my name was called for a visit with my physician.

My times spent in the waiting room, anxious as they were, became times of contemplation and introspection, of dealing with my own illness. I knew that only a few people live long and fulfilled lives without passing through a period of illness when health concerns and medical treatment take center stage in their lives. I had heard many elderly people talk about their illnesses. Most seniors know they will occur, although few plan for their arrival. The illnesses just appear at a time not of their choosing. Was I passing into my senior years?

And, how did one find meaning in ill health? That question was now on my doorstep. I couldn't help but think about the terrible illnesses that sometimes fall on young people, even small children. This, most of us believe, is a tragedy. A life waylaid, sometimes even snuffed out before it can get started. How does one find meaning in that?

Or, more pertinent for me at the time, how did I find meaning in this illness? While I knew that I might die from the disorder, it did not seem a real possibility. There were too many reasons not to die. I was not old enough. I had a job that I enjoyed. I liked working on my farm. I wanted to write more books, and my wife and I had several trips planned. Death was not a part of our game plan. So, why at a time when my life was filled with meaningful daily activity, did so much of my

Return to the Farm

attention have to be directed to medical concerns? There were no immediate, clear-cut answers.

The efforts to regain health were carried out in a twofold effort. First, I would follow rigidly the medical treatments as prescribed by the hematologist during those consultations back in the examination rooms of her office. Initially, I chose to take medications rather than turn immediately to surgery. Thus, for several weeks I took powerful steroids in the hope that the medication would kick-start my immune system back into action. Other potentially helpful and expensive medications, administered in the hospital, offered only temporary respite. Throughout the winter the platelet count fluctuated erratically, but always in an ever-decreasing downward spiral.

While driving home after one particularly powerful injection administered at the hospital, I experienced the on-set of a searing reaction, simultaneously experiencing a fever, disorientation, and a dry, hacking cough — necessitating a disoriented trip back to the emergency room for relief.

By themselves, the medications seemed ineffective. But I held out hope that surgery would not be necessary since I also attempted to pull myself back to health through spiritual exercises and alternative medicines. I read books about alternative medicine, reread the Psalms of David from the Old Testament, daily surrounded myself with new-age music, meditated, ate proper food, and rested. "Could not a body rightly aligned with the spiritual forces of the universe right itself," I wondered? My recovery became a personal test of my religious faith.

The church that I attended offered comfort. At my request a small group of us, eight in all, met at our small church for a prayer service that lasted no longer than thirty minutes. We lit a candle given to us a few days earlier by friends. Its small light shone brightly throughout the service. Our saintly

Doc Charley and other Doctors I have known

prayer partner who led the service had never before led a prayer meeting for the sick. I had never been the object of a prayer service. With absolutely no previous experience to go by, we created a religious service that was perfect for the occasion. She read scripture, several of us shared our thoughts, and she prayed. The few minutes passed quickly and the service was completed. I felt immediate relief. I resolved to make prayer and meditation forevermore a part of my daily routine.

But, relying on spiritual help was not easy, especially when physicians and family were suggesting surgery, and the dangerously low platelet counts fluttered even further downward. A fast recovery, such as might be possible with a surgery, would allow for a long-planned trip to Europe with my family to visit other family members; a slow recovery, with alternative approaches, meant disappointing others and staying at home. Meditation was followed with acupuncture treatments and massage therapy—both gave peace to the mind and stimulation to the body, but neither curtailed the decline in the platelet count. My hope in alternative cures faltered.

Finally, during a consultation in the hematologist's office in late winter, the lab report came back that my platelet count had again fallen to a precarious low. It was urgent, she believed, for me to take yet another series of injections to raise my platelet level, then immediately submit to surgery. Depressed, with erratic emotions and body bloated with the side-effects of medication, I could do nothing but submit to a questionable surgery.

Within days I was lying in a hospital room, waking up from surgery, my mouth cotton-dry, but able to distinguish the voices of my wife and friends. All, it seemed, had gone well and, a day later, the surgeon reported on his efforts. "We just sucked that spleen right out of you," the surgeon related as he described the surgical procedure he had used. "Your spleen

Return to the Farm

was larger than we thought it would be. The surgery was quite a challenge, but all went well," he continued, a touch of pride in his voice at having triumphed in a difficult surgery. Lab tests a few days later were even more encouraging. My platelet counts had returned to normal and have since so remained. I am healthy again.

But what does one learn when one is confronted with death and yet survives? I can report, like have so many other people, that each day is more precious and filled with more meaning.

Yet, thankful as I am for the surgery, I wanted to recover with the more unconventional treatments. I wanted to point to the fact that my religious beliefs, spiritual exercises, and non-traditional therapies were largely responsible for my becoming whole. But that was not to be. The slow curative restoration that I had expected did not materialize. In the end, surgery prevailed. I was left with the question of how to blend these two forces, religious belief and the practice of modern medicine! Without one, I would have no purpose. Without the other, I would have ceased to exist. What is healing and from whence does it come? The answer, for me, is yet to be determined.

B.F. and Kate's Jersey Cow

Our neighbors, B.F. and Kate Langston, bought a Jersey cow the other day. Today, that is of interest largely because it is an oddity. But sixty years ago when B.F. and Kate bought their first Jersey cow, it was a custom of these hills.

Back then, it meant that B.F. and his new bride, Kate, were settling into married life very nicely, thank you. The newly married couple purchased a Jersey cow so that they could have all the milk the two of them, and eventually, their children, could drink. Because Jerseys give milk rich in butterfat, Kate churned all the butter the family could spread on the biscuits she made for breakfast. They stored the milk and butter in the spring down the hill not far from their home.

But having a Jersey cow also meant that B.F. and Kate had a hog too—to feed the excess milk to. B.F. would take the milk out to the barnyard, holler "soo" and the little pigs would come running. In the fall, with acorns and field corn added to their diet, the pigs became the bacon and hams and pork chops that were salted away in the larder.

I remember when I was a kid, long before I met B.F. and Kate, and right after we moved to the farm, our neighbor from across the valley, George, a widower in his mid-fifties, married Rose, a spinster nearly his own age. They went out and bought their own Jersey cow.

On those mornings when we got up at daybreak, we could hear George's wail float across the valley, "Come Sally," and we knew that George was on his way to the barn to milk his Jersey cow. It was at Rose and George's house that I got my first introduction to hot home-made rolls smothered with real cow's butter. Rose also served bright red strawberry preserves, made from berries she had picked from her own garden. Sunday dinners were always a delight at Rose and George's house, thanks in part to their Jersey cow.

Return to the Farm

Rose and George also kept a calf or two penned behind a wooden framed fence in the corner of their barn lot. George would only milk enough for the couple's use, and then turn the calves loose. The two calves always knew where to find Sally, and would charge, bawling as they ran across the barn lot, then thrust their bobbing heads into her hindquarters and, with their tails wiggling to and fro, finish milking the cow. Thus, Rose and George always had a veal calf or two around the place to supply meat for their table.

We had Jersey milk cows on our own farm, too. When Dad started our dairy right after we moved to the farm, about half of the cows were Jerseys. I liked them best. They were gentle brown cows with a dipped forehead and large, brown eyes. They always looked and acted like ladies, I thought. For a time we milked our cow herd by hand, twice a day, until progress overtook us and Dad purchased a milking machine which made the task infinitely easier.

But of all the cows we had, the Jerseys gave the richest milk—their milk had the most cream in it. We would store our milk for home use in gallon glass jars in the refrigerator where the cream would rise to the top, leaving a golden layer of cream over nearly blue milk. The jar of cold milk always seemed to be covered with droplets of water, and the milk had to be stirred to remix the cream with the milk before we could drink it. But my, it was good.

Maybe the best time for eating was in the spring of the year when the strawberries were ripe. Several families always got together, usually at our home, and ate all the strawberry shortcake they could stuff down. Those neighbors who had strawberry patches at the ends of their gardens, including Rose and George, always brought gallons of fresh stemmed and quartered berries. Another neighborhood family brought homemade ice cream, made only from pure cream. Mother

B.F. and Kate's Jersey Cow

would always have white cake and whipped cream prepared. No one, even us children, were ever told we could not go back for seconds, or thirds, or fourths...

I even owned my own Jersey cow once. I named her Wanda, after a classmate at school. Wanda, the cow, was an FFA project, and one summer I took her to local fairs and even for one glorious week to the Ozark Empire Fair in Springfield when I showed her in a class of young Jersey heifers. She and I became good friends that summer.

Eventually Wanda became a productive member of our dairy herd. When I returned home from college on weekends, I would walk into our barn lot and Wanda would bawl and stroll over to me, demanding that I rub her behind her ears. What a nice welcome home.

Just the other day I saw B.F. and Kate's brown Jersey cow in the barn lot. She was docile-looking, with big brown eyes and a lady-like structure. I don't think B.F. purchased her because he needed to save money on the milk he and Kate use. I know that he recently rented several hundred acres of Stone County pasture land and purchased another hundred or so beef cattle to go with the herd he already had. He doesn't act like someone who is going to starve to death unless he produces his own milk. But I do know that the milk tastes good. I drank some of it the other day sitting in their kitchen and then watched as Kate finished churning a pound of butter. She carefully salted, then patted the fresh butter into a pound chunk before placing it in her refrigerator. It sure looked good. I think the reason B.F. got that cow has something to do with the quality of Ozark life.

Yet, keeping that old cow is not easy. I can imagine B.F. getting up in the pitch darkness of morning, calling his cow as he goes to the barn and sitting on a T-stool in the glow of an electric light bulb, his head pushed into the flank of the

Return to the Farm

cow, yanking away, and hoping her dirty tail does not rake him across the face. That's the way it has been in the Ozarks for a long time. B.F. and Kate carry on that tradition.

The Pecan Orchard

The pecan orchard came into existence a few months after Dad died. One day as we were talking, Mother said, "If you know of any additional uses for the farm, please let me know. We can talk about it." Since I had always believed a farm should be diversified in its crops, I suggested, "Let's plant a pecan orchard. It's still a few years before I retire. We can plant them now and by the time I retire, they will produce a small, supplemental income and also give me something to do. The field between the railroad track and the creek would be ideal. The soil is excellent. Water is available."

I had visions of magnificent pecan trees growing across the bottomland next to the creek. In my mind I could see the pecans setting on in the spring, maturing during the summer months, and then falling to the ground with the first frost of the fall. That's when I would need to gather children and grandchildren together and we would go out into the field and spend a weekend picking up these tasty morsels of food, crack some open to eat on the spot, and bag the rest to be sold to neighbors and friends, or carted off to market. That would be a nice way to spend part of my retirement years.

Boy, did I have a lot to learn.

At the time I didn't know much about growing pecans, only that I liked trees. And, I knew that pecan trees, while normally thought of as a southern tree, grow as far north as the Missouri River, several hundred miles north of our farm. We could grow large native nut varieties which, while not as large as nut varieties in the southern part of the United States, have superior flavor. We would grow pecans that had an excellent flavor and eye appeal.

But Mother objected to my suggestion, even if ever so slightly. "No," she said, "that will not work. That field you've suggested has the best soil on the farm. We need it for hay."

Return to the Farm

For a moment I was perplexed. Hadn't she just asked for my suggestion? "Is it the pecans you object to, or the field we are using?" I asked.

"It's the field!" she answered.

"Then we could use the field on the other side of the railroad track. That field is similar in size and the soil is equally rich. Can we use it?" I asked.

"That would be acceptable," she said. And so Karlene and I began to make plans to plant pecan trees.

Mature pecan trees are huge trees that tower into the sky, their limbs surging upward, producing clusters of nuts high in the tree. Like any other farm crop, the trees need care and considerable effort from a grower before the nuts, which vary in productivity from year to year, can be harvested. I learned from my reading that as much work would be required in raising pecans as in any other commodity. Yet, the venture seemed to suit me.

An immense amount of work is necessary to start a pecan orchard. Ten acres, or approximately 300 trees, was large to me even if a minuscule number when compared to the huge plantations comprising hundreds and sometimes even thousands of acres in the southern states of Georgia, Louisiana, Texas, and New Mexico.

We needed first to "lay out the field," that is, decide where each tree would be planted. Because pecans easily grow to heights of over 100 feet, mature trees are normally thought to be most productive when they are located about seventy feet apart—in a crisscross, so that you can look down a row either way and see huge trees growing in a straight line.

The problem is that it takes at least fifty years to grow a tree that will be over 100 feet tall and spread seventy feet across; in the meantime, a lot of space between trees is not productive, not being used. A tree should begin production when seven to ten years old, and by fourteen years of age should be into good production.

The Pecan Orchard

Pecan orchardists learned long ago to plant trees thirty-five feet apart, both ways in the row, and then, after twenty-five years or so, to begin thinning them by removing alternate trees so the remaining trees can reach full maturity. We decided on this method.

I was discussing my desire to plant a pecan orchard with a member of the nut growers association meeting one day—I was in my early fifties at the time—and he responded, "That's a good idea." He was considerably older than me, and a person who knew and loved his trees. "But you know, young man," he responded, "people who plant pecan trees are usually much older than you. You're getting a head start on your trees. You'll probably live to see your trees produce." His comments spurred my resolve to have an excellent orchard.

Karlene and our grandchild, Afton, helped lay out the field for planting in what is probably the most difficult task I've encountered in establishing the pecan orchard. The field would best allow, I thought, for rows ten trees wide and thirty-one trees long. Since trees could not be planted in a small bog near the center of the field, we would have roughly 300 trees, plenty to care for in my retirement years.

The latter part of February we were ready to plant and so we crossed the rectangular field with twine, tying a white rag onto the twine every thirty-five feet so we could lay out our ten rows. Then, as we finished a row, we would move the twine down the field exactly thirty-five feet. The system worked. At the end of the day newly planted trees stood upright in fairly straight lines.

The trees came from two different sources—I had ordered some trees from the Missouri Department of Conservation, which sent me several bundles of twenty-five trees each. The spindly one-year-old seedlings were healthy but not over a foot tall. "They have a long way to go before they tower 100 feet into the air," I thought.

But since I had not purchased enough trees for the field, I also obtained a brown grocery sack of pecans to use as seeds to plant the remaining part of the field. The seeds had come from a pecan orchard owner north of us, and he had striated the nuts—frozen them for 30 days—to make certain they would germinate. They would, I was told, grow into trees faster than the seedling trees since they would never have to be transplanted nor would their root system have to adapt to a different soil. To plant the nuts, we carefully sat a coffee can, with both top and bottom removed, on the spot where we wanted the tree to grow. The can would protect the pecans from rats, mice, or birds that might want to eat the nuts. Next, we planted three pecans in each can and covered them with soil. Later, when the trees began to grow, we would remove all but the most vigorous tree, and our orchard would be established.

We were thankful for our granddaughter, Afton, that day. We needed twine, we needed trees, we needed pecans, and we needed cans all day long and clear across the field as we worked our way from one end of the field to the other. She had all the energy of the seven-year-old that she was, and she talked, asked questions, and brought us the items we requested, saving us many steps. We all slept well that night.

Another orchardist talked me into planting a dozen black walnut trees at the west end of the field, and the orchard was set.

Tender young leaves began to emerge on the trees around the first of May. Slowly, over a period of weeks, one tree, then another, would produce the tiny leaves, which showed that the seedlings had taken root and were alive. The germination of the pecan nuts was much the same, although

The Pecan Orchard

they were slower to emerge than the tiny leaves on the seedling trees. By the first of June I could peer into the coffee cans that harbored the seeds and begin to find the first of the trees poking a bowed stem from the soil—the first growth of a trunk that would eventually tower 100 feet into the air. By July, most of the plantings had a sprig of green—an indication of life. Our effort was successful. Those who told me that the nuts would produce a tree as fast as the seedlings proved to be correct. The small seedling trees, although most of them lived, grew very little the first year. They just stood in the field holding onto the tender young leaves, but doing little more. They seemed to be adjusting to the soil in which they had been planted. But the trees from the pecan nuts, although they emerged from the soil weeks after the seedlings broke bud, continued to grow as wiry trees until, by the end of the season, many were as tall as the seedlings.

At first I thought I would have an organic orchard—that I would use no pesticides and no fertilizers. I purchased a good hoe to use on the weeds and a small, mechanical weed cutter. So armed, I attempted to keep all weeds at least three feet away from each growing pecan tree. I soon learned that my efforts were totally inadequate. I had a limited amount of time and the grass and the weeds grew at a rapid rate, even when, in summer, a small drought occurred. That first year I even began watering each of the plants by hand, with a bucket and dipper, in an effort to encourage their growth. But this effort too, even though I worked in the hottest heat of the summer, proved to be inadequate. I was ill prepared to water 300 trees with the meager tools and machinery I had. Before summer's end, the trees were destined to stand before the forces of Mother Nature endowed only with the hereditary factors that nature gives to all trees. But nature provided adequately.

Return to the Farm

By fall, in among the weeds and fescue grass, I discovered that most pecan trees were at least a foot high. Difficult as it was, our fledgling orchard was taking hold.

I took little notice of the small trees that first winter, believing that I needed only to do a much better job ridding the trees of weeds in the coming year. It would be spring before I discovered that the small trees were a bonanza for the rabbits, rats, and mice that lived in the field.

That winter a heavy winter storm left the field crusted with snow and ice for a couple of weeks. During the time the field was covered with the blanket of white snow, hungry rabbits, rats, and mice discovered the fledgling pecan trees. The rodents must have thought them dainty morsels for, when I walked across the field in the spring, I found many of the trees not just eaten on, but sometimes even girdled and on occasion completely eaten through and lying on the ground. These critters of winter had done more damage to the field than the drought of summer.

For the next several winters, we waged a wintertime war against the rodents. Each spring I purchased enough small trees to replace the ones that had died, then obtained a small protective sheaf to fit over each tree—a day's task just going through the field and placing them on the trees. To tidy up the field, we asked our neighbor to bale the hay between the rows, then threatened him with his life if in the process he inadvertently cut even one small tree. Finally, we started allowing the cows to graze in the orchard, but in order to protect the trees, only after the first heavy frost of the year knocked the leaves off the trees. The grazing of the herd shortened the grass, making rodents more visible to the watchful eyes of hawks and owls.

I also began spraying the weeds and grass around the trees with herbicides. The regimen began to work. Within

The Pecan Orchard

three years, the trees were standing out over the tops of the weeds and grass. You could actually stand at one end of the field and count the thirty trees growing in a row, clear to the other end of the field. I was beginning to feel that the orchard was really an established orchard, and that it would eventually mature.

Next, I knew, I would need to graft every tree in the orchard to make certain that the trees would produce a genetically hardy and productive nut suitable to our region.

But, first, I had another huge lesson to learn. I did something that can only be described as stupid—idiotic—when I broke the one absolute rule that I know about fires. I know never to leave a fire unattended, but I did anyway. The results were catastrophic. Still to this day, it is difficult for me not to castigate myself for what I did.

It happened this way. Along one side of the pecan field, about halfway down the field and next to the railroad tracks, three old, mammoth, and dead sycamore trees stood, shedding their gnarly, twisted branches. On one occasion I spent much of an afternoon watching nesting wood ducks entering and then leaving a large hole at the top of one of the trees. But these trees were becoming a nuisance, perhaps even dangerous. Their uppermost limbs had broken during previous storms and were lying tangled in a pile at the base of the trees. Perhaps a limb would fall on a cow during some storm, or the trunk would fall on the fence, obliterating it and allowing animals onto the railroad. The limbs, and the three trunks that towered skyward, needed to be cleaned up and burned so that the field would look neat and cared for. Burning those trees became a housekeeping duty.

In the early morning one February day I lit a fire to the limbs at the base of the trees. My efforts were immediately rewarded. Soon a crackling fire was consuming the limbs and

Return to the Farm

licking away at the base of the three dead tree trunks. They too, would soon catch fire and burn. All afternoon the fire continued to burn. At times it would bellow, then nearly extinguish itself and I would have to pile more limbs onto the flaming ashes to make the fire crackle again.

The tree trunks, however, did not burn so well. They were mammoth, and while they were dead, the massive insides were still partially green and wet. The fire, while it seethed and smoked within the trunks, could not wholly burn the cavern. The smoke wafting out the top of the trees indicated that they were smoldering, but it was impossible to tell how far up they burned. Probably, I thought, they would burn themselves out.

Late in the afternoon I raked the coals at the base of the tree together, then sat and watched as they burned almost to extinction. Wishing I could put out the invisible fire that smoldered in the tree trunks, I wondered about going home. There was no wind. It would, I finally concluded, be safe to leave. I would not stay any longer and watch the fire. I left for home.

By the time I reached my house an hour and half later, I found a message on the answering machine from my sister. The grass in the pecan orchard was ablaze. The fire had traveled eastward across the orchard, burning grass, trees, a boundary fence, and eventually some of my neighbor's field. The neighbors and the local fire department had all descended on the field. The fire was almost out. Many people were still working on the fire. Didn't I want to come and help?

My son, Chris, who arrived for a rare visit just as I was leaving for the farm, accompanied me. The wind had come up after I left, and sparks from the massive tree trunks had blown embers onto the field. Two-thirds of the eastern side of the field had burned. The damage to the trees did not look too significant. Some trees had been badly burned, but others seemed only to be faintly scorched by the heat. I apologized to the

The Pecan Orchard

neighbors. I felt that most of the pecan orchard would survive intact.

The sycamore stumps were still standing and smoldering. No one had been able to get them out. Chris and I spent much of the night trying to get additional water into the still massive stumps until all signs of smoke stopped. We thought, finally, the fire in the huge sycamores was out, but we could not be certain. We could get no more water into the trees. It was almost daylight when we left the field.

But this effort was not the end of the fire. Three days later the wind again began to blow and shifted a full 90 degrees, this time blowing due north. By then, we later learned, the sycamores were again spewing embers. A few of the embers blew across the railroad tracks and landed in still another field covered with dry grass. Again neighbors and members of the fire department raced to my field. The fire was not so extensive this time, but my mortification reached new heights.

Though by this time nothing was left around the sycamore tree trunks to burn and they continued to smolder for a few more days before falling over and going completely out.

In the spring I learned that the pecan trees that I thought had only been lightly scorched were, indeed, dead. Not only did I lose two-thirds of the pecan orchard to fire, but because I did not replant them immediately, thinking they would rebound, I lost a full year of growth on any replacement trees I would plant.

My retirement income had taken a severe beating.

Still, we decided to replant the east two-thirds of the field that had been lost to fire—approximately 200 trees. It was easier to replant this time since stubs of the burnt trees were still sticking their heads above ground and served as a guide so we did not have to align the new plants. The young

Return to the Farm

trees arrived the latter part of March and I took a shovel to the field, made a small slit in the ground where each new seedling was to be planted, dropped the seedling in, and removed the shovel. The planting went quickly. But it was discouraging since we were starting over with so much of the orchard.

The trees grew much as they had the first time. Most took hold that year although, as before, for the next three years I had to replant those lost to mice, rabbits, and hot weather.

I had learned something about establishing trees. This time I did not even attempt the primitive and arduous task of watering the trees by hand in the hot summertime. The small trees were on their own, even in 100-degree heat of summer. But the trees were hardy. Very few died. I did work harder at killing weeds and fescue grass around the trees—although this time I left my hoe and cultivator at home and used chemicals. It was so much easier to place an old stovepipe over the small trees to protect them from the spray and then, using a hand sprayer, drench a three-foot circle with herbicide/water solution around the trees. Two such sprayings a year and the weeds were reasonably well controlled.

Excitement developed in the fall of 1993 when a flood covered the field. This was the same year of the big spring flood through the Midwest and down the Mississippi; except our flood, of much smaller dimension, took place in the fall. In just a very few hours a twelve-inch rain fell in the area. Much of the rain fell above our farm, and the water flooded down Clear Creek as it has on very few occasions. Crops, cows, and pigs all floated down the surging creek. The pecan orchard was inundated with water that I'm sure was well over waist deep. Our next-door neighbor, who raises gourds for the commercial market, could only watch in dismay as thousands of his bobbing gourds floated across the orchard and down the valley, disappearing around the bend in what was now a small stream

The Pecan Orchard

turned churning river. But the flood did very little permanent damage to the pecans and my neighbor salvaged enough of his gourds to stave off a financial disaster.

Within a couple of years, I had a two-tiered field—on the east one-third of the field where there had been no fire, the trees stood head high—on the west end the trees stood little higher than the waist-high summer fescue grass.

My retirement plan, delayed, was back on track.

One major task still had to be accomplished with the pecans before production started—the immensity of which I had not anticipated. Pecan trees need to be grafted.

As one of my friends stated, "You never know about the genetics of the pecan tree you plant, whether or not it is going to be a high producing tree." High production can only be ensured when a small twig is cut from a high producing tree and bound—grafted—to the growing seedling in about the third year of its life. That graft becomes the central trunk to the tree.

Skilled professionals, those old orchardists who have grafted trees for decades, can complete a graft in a few minutes. For me it took much longer. I attended an afternoon grafting school, purchased suitable seine or graft wood which was cut in late winter, and kept it cool in the refrigerator until the first of May when the sap was beginning to rise in the new seedlings, making conditions right for the actual grafting of the tree.

The first year I grafted about twenty trees and half the grafts "took." The second year both the number of trees I grafted, about fifty, and the percentage of "takes" rose. I was gaining confidence in my ability by now, so the third year I grafted one hundred trees—but had a miserable twenty percent success rate. My confidence sank until my friend Gerald Gardner, a long-time nut grower, came to my rescue. He came to the orchard one warm May day and gave me an in-field

Return to the Farm

demonstration, grafting thirty trees which achieved 100% success. I was more than impressed with his abilities, I marveled at them. How could anyone achieve such success? But, he did!

But alas, the following year, overflowing again with confidence, I undertook to graft 200 pecan trees. The effort took me a full week as I had to secure the seine wood, prepare the material, and then spend several days carefully cutting trees, selecting and shaping the seine wood, and finally binding the new stem to the growing tree. It was with great expectation of finding many newly formed buds emerging from the many grafts that I approached the field many months later. But, it was not to be. I could find only a precious few of the new grafts breaking forth with fragile green leaves. The remaining grafts remained dormant, just as I had bound them to the tree. My success rate, I estimated at the time, was less than ten percent.

The following year Gerald came to my rescue again, this time bringing with him his friend and fellow pecan grower, Doc Roberts from Joplin. The two arrived at the pecan field early in the morning with adequate amounts of seine wood in their ice chests to last all day. Then, with the skill of surgeons they spent the day grafting—preparing the live tree, then selecting, shaping, and binding the proper piece of seine wood to each tree, all the while giving me a refresher course in the art of grafting. Together, we grafted a hundred trees in that single day. A month later, when I revisited the orchard, I found the "take" ratio of the previous year reversed. Through their efforts, 90% of the new grafts had attached themselves to the saplings. My confidence and belief in having control of the orchard again returned. Next year, I thought I should be able to complete the grafting of the orchard.

The trees grafted the first year are already flourishing, the most vigorous among them perhaps twenty feet tall with

The Pecan Orchard

trunks as big around as my doubled-up fist. The faster-growing black walnut trees in the orchard, although much fewer in number than the pecans, are already setting their first crop, and I expect next year to gather nuts from the pecan trees. After that, each year for many years, the size of the crop should increase.

Often when I am at the farm, I go down to the orchard and walk from one end to the other, observing the trees, checking for growth, damage from the elements, or rodent destruction. But, mostly, I am pleased and encouraged that a project that has taken so long to unfold seems destined for success.

Others share this interest with me. Many of my friends often ask about the pecan orchard. "How are the trees doing? Do you have your first pecan yet?" they want to know.

The pecan orchard holds a special place of interest for Karlene and me. We planted the orchard together and have watched it grow, and we both anticipate her baking that first pecan pie. One evening last year, in an enthusiastic and yet introspective moment, I remarked that, when my days have passed, I want my body to be cremated. Then, the family, if they want, should have a party at the west end of the pecan orchard, tell stories and laugh, and cry if they must, eat a piece of pecan pie, and then, catch a strong west wind and throw my ashes into the air. In retrospect, I still believe the orchard would be an appropriate final resting place.

But, I hope that day is long into the future.

I know now that there will not be any "easy" retirement money from the orchard. Pecan production, like all other farm enterprises, requires hard work. The work requirements evolve over the years, from planting and weeding and grafting when the trees are small to examining larger trees for insects and disease and controlling damage and finally, when the trees are grown, to harvesting and marketing the nuts. Like the rest

Return to the Farm

of farming, to skimp or take a shortcut in the orchard is to be shortsighted.

Now my enjoyment comes from watching seedling trees take hold in the earth and grow—usually about a foot a year. It's a slow growth with a payoff well into the future. The major production of the orchard will undoubtedly peak years after I am gone, perhaps even after the planter's identity has been forgotten.

I've had some special moments in the field, perhaps the most memorable being the early spring day when I noticed a bald eagle, with white head and tail feathers reflecting in the sun, flying a couple of hundred feet in the air toward me from Clear Creek, then turning as it passed directly overhead, as if satisfied with what it had seen, and returning in the direction from which it had come. A coincidence that its path was directly overhead? I prefer to think not. It was a good omen for the pecan orchard.

Rarely do I go to the farm but that I eventually visit the pecan field and walk up one row and down the next, examining the condition of the trees. By the middle of May their tender green leaves have normally opened, by the middle of July they have grown a foot or more, in August they may be showing signs of inadequate water—a drooping plant or wilting leaves. This is when I wish I could irrigate them. By late fall a hard freeze has trimmed each leaf from the tree and I can turn my cattle into the orchard, to feed on the remaining fescue grass still on the ground. It's become a cycle and one I have become a part of.

I find that tending to the trees is a reflection of my confidence in the future. People will always need food, and pecans are a special kind of product which people can enjoy.

I look forward to my first pecan pie, baked by my wife from the pecans we have harvested from our own trees. It will

The Pecan Orchard

be a major celebration and I hope many children and friends will be with us to help eat that first pie.

Note #1

I was surprised recently while reading an article in a pecan producer's publication. The author of the article shared her opinion that both beef and pecan nuts supply protein, but that beef production is an inefficient method of supplying protein since normally enormous amounts of corn are necessary to produce most of the commercial beef we eat. She asserted that most people include too much beef in their diets. She was adamant in stating that protein in pecan nuts is equally healthy for the body and much less costly to produce than beef since it takes more resources to raise a pound of beef than a pound of pecans. She hoped that not too far in the future Americans would substitute nuts, or more specifically, pecans, for beef protein. I wondered, could such a substitution make for healthier bodies at less cost? Perhaps, inadvertently, by planting a pecan orchard, I helped supplant the very beef industry that now supports my farm.

Note #2

Yes, yes, yes! Now, it's the Fall of 2003! For naught I looked up into my pecan trees this summer, hoping to find my very first pecan growing from bud to maturity. But, at no time did I find a pecan on any tree until last Saturday, when returning from a pecan producers meeting near Joplin, I made one final examination of the orchard. To my utter surprise, there at the top of one of my largest pecan trees hung two large pecans, my first. My eyes widened, my heart raced, I shrieked for joy.

Return to the Farm

The pecans were high in the tree. I could not reach them but was able to get a long stick and, after several tries at whopping at one of the poor little pecans, dislodged it with a mighty whack that sent it sailing earthbound and into the weeds. I was never able to find the fallen pecan. Not wanting to risk losing the second pecan, I hurried to the house, found a ladder, and returned to the tree. Soon, fifteen feet above the ground, I picked my first pecan.

The Year 2000

I was glad I owned a small farm when the year 2000 rolled around. True, I got caught up in the frenzy about computer failure. How could this modern world be both so complex and so fragile that concerns about computer programming failures brought fears that the world would be thrown into anarchy? Many people were concerned about energy and transportation disruptions. I was concerned about food disruptions. Would we have a great famine? Would hunger erupt because our mechanized world would no longer be able to produce and deliver food?

I was made aware of the implication of the computer breakdown when a friend of mine at work began alerting me to the potential problems. This was months before the press took up the issue. My friend was really concerned.

He began telling me what the news media would report months later—that computers were ill programmed. He feared they would not recognize the new millennium. The dating system of home computers, he told me, would probably revert to the previous century, then freeze up. Others would just freeze up. Families would not be able to use their computers. When this malfunctioning technology was multiplied to large, multinational corporations with their vast networks of computers, his account of impending disaster was horrendous. He, in fact, seriously considered quitting his job and spending several months preparing for the approaching disaster. He felt this might be the most responsible decision he could make—and he wondered why so few other people were even considering the choice.

His description of the coming doom caught my attention. I tend to be a little on the pessimistic side of most issues anyway, and his logic made sense to me, especially after the media began sensationalizing the same sentiments.

Return to the Farm

The two of us discussed the issue fervently at work and with anyone we could find who did not think we were too "cuckoo." (It was only with those who obviously considered us too far out on the fringe that we would shut up. No, we didn't want people to think we were too abnormal.) But, we did want to take care of our loved ones, even if it meant taking drastic actions.

As the weeks passed and the fateful day steadily approached for the impending disaster, I had to decide what, if anything, to do. I believed there would be some disruption to the economy, some computers somewhere throughout the world would cease to function, and that somehow a mess would develop that would, in some way, eventually affect our family. Somewhere the food supply would probably be disrupted, I decided. I also decided—not that I needed to go anywhere—that I would not get caught riding in an airplane the night of December 31, 1999. In fact, I would be staying home that night.

But what to do? How do you prepare for such an eventuality? I felt, first, that if the disaster occurred, we could not rely on local supermarkets for food because most food is not raised locally. Many of the eighteen-wheelers that whiz past each other along the highways are carrying food—oranges and vegetables from California and Florida, beef from Kansas, corn and soybeans from the Midwest, milk and cheese from Wisconsin, and chickens from Arkansas. I feared that these trucks might stop moving and our food supply would be shut down.

Many of the news commentators I listened to indicated that people should, at a minimum, purchase an additional two-month supply of food. An additional two-month food supply, they reasoned, could get a family through until the government

The Year 2000

functioned smoothly again and food supplies became readily available.

I liked the idea. Enough food to last an extra two months seemed reasonable to me.

Late in the fall when I was in Springfield, I went to a cut-rate grocery store and got not one, but two of their largest grocery carts and walked through the store, aisle by aisle, carefully noting the huge number of products the large store had to offer. By the time I was at the checkout stand, cans, boxes, yes, cases of food were piled to overflowing in the carts. The two carts barely held all that I thought we needed. Surely this was enough food to last two people two months, with some extra food to share with our children, if they needed the help. I thought it would be enough to get us through any food disruption that might occur.

And, of course, I was keenly aware of my little farm. We raised little edible food for ourselves—no garden, no fruit trees, only cattle which we sold. But, the soil is very fertile. We could raise food, lots of it, for our own immediate consumption and the consumption of others.

So, I made one more stop on the way home—at a garden supply store. I discovered that in late November last year's garden seed was on a half-price sale. "Just in case this shortage lasts longer than two months," I told myself, as I purchased enough half-priced corn and peas and beans and cabbage and broccoli seeds—and whatever other kind of half-priced garden seeds to produce enough food for two people, and their children and grandchildren—to last a year. We could butcher an extra beef, too, I thought, if we needed additional meat. This was our insurance against a worldwide computer disruption.

I stayed up late on the night of Dec. 31, 1999 and watched television productions from around the world and was amazed to discover that today's television producers are capable

Return to the Farm

of producing good television after all. I was overjoyed to witness the many ways different countries throughout the world celebrated the new millennium. That was the first good omen of the evening. The second one, of course, occurred when airplanes did not fall from the sky, utilities did not shut down, and major food supplies and other commodities were not disrupted. In fact, I was relieved (along with millions of others) to learn that throughout the world very few disruptions occurred. The dire predictions had been for naught.

(In retrospect, I now like what the Italians had to say about the arrival of the new millennium. "We had no problems with the arrival of the first two millennia so why should we be concerned about this one?")

But this, of course, is not the end of the story. As we all know, on September 11 our way of looking at the world changed forever. Although we had repeated warnings of the possibility of such attacks, it was not until we witnessed the sight on TV of airplanes, human-guided bombs, deliberately flown into our national buildings did we mobilize for disruptions to our way of life. The actions of the day seared into our consciousness the knowledge that we are not invincible. On this occasion, the immediate aftermath brought a disruption to our airline system.

Government officials now tell us that the threat of other disruptions is continuous. We are left only to ask when, eventually, will terrorists strike again or a rogue nation use a weapon of mass destruction against us? Or, even from within, could there not be bombings from disenfranchised militant groups? Other kinds of major calamities are possible: a rippling California earthquake, a national fuel shortage, an electrical blackout, or a crippling computer virus; even a major transportation breakdown, for whatever reason, would have calamitous effects.

The Year 2000

Of course it is impossible to prepare for all eventualities. We're all human. But even before September 11, experts within the ranks of our nation's internal police forces were warning us about the lethargic nature of our intelligence forces. Critics explained that these agencies were dominated by their own bureaucratic restraints and self-imposed gridlock. While the agencies seem to have righted themselves with remarkable speed, sometimes even to excessive action as they look to repelling the potential invaders from within, even their most ardent supporters admit it will be years before they can form a fully-functioning intelligence community. And, does anyone believe that even if successfully formed, they can stop all such attacks forever? And, of course, even less can be done to prevent debilitating natural disasters.

This fear ignites my concern that when the additional disruptions occur, people will go hungry. With all of our recent concern about the welfare of the nation, I have heard almost nothing about ensuring a continuous food supply during times of disruptions. Wouldn't it make sense to develop a national plan or policy to ensure, insofar as possible, that food would be available to all, in sufficient quantity and quality to provide an adequate diet, during the time of a national disaster?

But, there are some real problems with ensuring this food supply. One is the make-up of our national farming community. Our nation was formed at a time when agriculture dominated American life. Farms were small; the farmwork performed by the hand labor of large families or with draft animals. When a piece of land was "used up"—the soil depleted of its ability to raise crops—the farmer simply moved west to homestead another farm. But, over the years as farm ownership became entrenched and landowners learned the art of farming and mechanized food production, farm ownership changed dramatically with the advent of the industrial revolution. New farmland dwindled quickly

Return to the Farm

until it was completely gone, and farmers learned that, through stringent farming practices, they could stay on one piece of land and continuously produce commodities.

Many changes were spurred by the United States Department of Agriculture. First, the Department of Agriculture established agricultural research universities, or Land Grant Universities, throughout the United States to develop and test the most scientific agricultural data concerning plant and animal production and viable farm practices. Next, a Federal Extension Service, with divisions in every state, was established to disseminate this information about plant genetics, hybrid seeds, fertilizers, mechanization, and selective animal breeding to farmers throughout the land. And, when farmers were too slow to adopt the the new scientific information, youth organizations, named 4-H Clubs, were established to further spread the information to the young, who then became the early adaptors of the scientific agricultural information.

The government-initiated research urged farmers to become more efficient, to specialize and mechanize. Farmers were told to concentrate their production and sell their produce to food wholesalers who would process milk and beef and vegetables into packages of palatable food to be sold to housewives in the grocery stores across America. The idea worked, insofar, at least, as producing sufficient amounts of food. America has become world renowned for the abundance of food produced by its farmers. In fact, America produces more food than Americans can consume, and a surplus of farm commodities—wheat, cheese, and corn products—are stored in warehouses across the nation. But, as might be expected, as production increased, prices for farm commodities decreased, often to the point that farming was no longer a profitable enterprise. To make money, farmers were encouraged to increase the quantity of their crops and livestock, to raise and sell more produce,

The Year 2000

even at the lower prices, in order to pay their bills and still live a comfortable life. To accomplish this most farmers had to purchase additional machinery and land, and take on additional debt. Thus did a vicious cycle of removing farmers from the land evolve. For, with prices of farm commodities heading downward, some farmers could not stand up to the pressure of the additional debt. Farmers sold their farms to other farmers and moved to the cities; the remaining farmers consolidated farms into larger tracts.

To rectify the situation, the Department of Agriculture advocated that farmers be paid a subsidy for their produce and on occasion, even paid for not producing. But the government subsidies promoted ever-increasing production and the downward spiraling of prices which, in turn, required farmers seeking efficiency to purchase more land and machinery, and contract for more debt. Contrary to the intent of the subsidies, more farmers could not compete with the efficiency of their neighbors and so even more farmers sold their land to the larger farmers and headed for the city until, today, America's production comes from highly efficient, specialized, and subsidized megafarms. Subsidies fed the spiral they were intended to ameliorate. (I keep remembering that when Dad had thirteen cows, we used to milk the animals in a half hour, and the sale of the milk contributed significantly to our family income. Now, we hear of dairy operations with over ten thousand animals where the mechanized milking process never stops.)

According to the census, in 1950, about the time Dad started farming, there were 5,722,500 farms in the United States. At the turn of the century, in 1900, 6,500,000 farms existed. Today, the census counts only 1,190,510 farms in the United States, with the count still going down each year. Expectations are that even fewer farmers will exist next year. And, with bio-engineered food now appearing in supermarkets,

Return to the Farm

the decline may continue for years to come—leaving fewer and fewer people with the knowledge to raise food.

The country is vulnerable to a second problem, concomitant with the first. The behemoth farms have not only concentrated on a small number of products and crops, but have adapted their crops to local conditions rather than diversifying.

In one part of the country we have huge wheat farms, in another huge corn and soybean operations, citrus growers in another, poultry production in Arkansas and on the East Coast, gigantic hog operations in North Carolina, concentrated dairies around metropolitan areas, and beef production in the South, Midwest, and West. Vegetables come from Florida and California. Under our current system, if Americans are to eat a balanced diet, their food must be transported long distances. Again, many of the huge eighteen-wheeler trucks we see on the highways whizzing past each other from coast to coast, north to south, are transporting food. And, in addition to the vulnerability of this food while in transport, other disadvantages to the consumer result from this system of food distribution. Perishable food selected to be transported must be durable, capable of keeping color and texture for long lengths of time—and this is often accomplished at the expense of taste, quality, and nutritive value. Tomatoes, corn, and beans produced locally for immediate consumption simply taste better. The energy cost of transporting this produce adds substantially to the total cost of the food in the supermarket. And, when one adds the fact that the highways and transportation systems could well become inoperable during periods of great calamity, one must conclude that our food production and distribution system is fragile and precarious at best, and our food supply is vulnerable.

Would not the welfare of the nation be better served if more, not fewer people knew how to raise food, and would it

The Year 2000

not be better if that food was raised closer to where it is consumed? And, would not, in all probability, the quality and taste of that food exceed that which is presently available and, because of lower transportation costs, be cheaper to purchase?

Huge farms, while dominating the agricultural sector, have not completely engulfed all small and family farms. Some smaller and financially successful farms still continue in operation. The Amish farms, for example, about which much has been written, continue as viable examples of successful farms. The Amish follow a pattern distinctly different from the intense farming practices associated with large landholdings and high machinery investment. The Amish, who have lived in this country for centuries, farm with a "low-input" system, making their major contribution to the farming operation through their own hand labor and, for many, horse power. Combined with their knowledge of agriculture, they expect each family member to contribute significantly to the operation of the farm.

The families on these farms are much more connected to direct markets, selling their produce directly to consumers or at farmers' markets, rather than to wholesalers and food processors who then transport the produce to faraway markets. Their farms are also much more likely to be diverse in nature, raising not one but many crops. And, although their farming practices run contrary to the recommended practices of the Department of Agriculture, the Amish are economically successful. Few Amish farms declare bankruptcy or liquidate their holdings nor, amazingly, do they take government farm subsidies. They thrive largely by producing quality food for their local areas.

Alternatives to the current agricultural practices of large farms, huge government expenditures, and artificially-induced production exist. It is not wise for our nation to follow a policy which ensures that each year more farmers are

Return to the Farm

wrenched from the land. On the contrary, I think the government should encourage more and more people to raise food, to offer assistance that helps sustain family farmers, and to promote smaller and more diverse farms rather than the larger and more concentrated agricultural operations that now dominate. But to do so would require a reversal of our farm policy. We would need to look at agriculture in a different light. But, promoting local food and smaller farms would be a far-sighted action to look after the welfare of all people. It could save more than just a lot of heartache in the event of another national calamity.

Berwick Community Church

When I work on the farm, I can look up the hill and see the Berwick Community Church—not a fourth of a mile up the road—that we often attended when I was a boy. It took a while for us to begin attending the church after we moved into the Berwick community. Mother was always particular about where we attended church—she had been raised a Baptist and pointedly wanted her children to be Baptists, that is to embrace Christianity as Baptist doctrine dictated. She even wanted my father, a non-Baptist when the two were married, to become a Baptist. I don't think she believed he was a heathen, just a non-Baptist. As a child I remember him being baptized by immersion, making him eligible to become a member of the Baptist church, but that was before we moved to the farm.

Mother made her torturous soul-searching decision allowing us to attend a non-Baptist church about a year before I graduated from high school, so it was only for that short period of time that I regularly attended services at the Berwick Community Church. I don't remember that I ever formally joined the church although I'm sure that Mother and Dad did at some time. (Nor, am I sure that she became a Presbyterian. It may well be that she just joined the church and converted everyone in it to the Baptist doctrine of faith.) At any rate, the last year I was home before heading off to college, we attended the Berwick Cumberland Presbyterian Church just up the road from the house we lived in.

The congregation dates back to 1842 when the earliest settlers came into the Clear Creek Valley. Church records indicate that a John Ritchey established three churches in the western part of the county, one at Newtonia, a second at the small town of Ritchey, and the other in Shoal Creek Valley. It is this last congregation which, before the Civil War, moved a few miles over into the Clear Creek Valley, the present location of

Return to the Farm

the church. At first, the congregation met in the homes of church members. But eventually, members constructed a building to house services. Of the three churches, only Berwick remains—traces of the other two having totally disappeared.

The church has been in near-continuous operation since its formation. Church records indicate that no services were held because of "Lincoln-madness" during the Civil War, and the church remained closed for four years, 1861-1865. Then, after the war, many members did not return or, because of their allegiance during the war, were not made welcome when services began again.

When the railroad came through in the 1870s, the church undoubtedly got a great boost for it added the name Berwick, taking the name from the small railway station and new settlement developing just a mile and a half away.

The current sanctuary, built of field sandstone, was completed in 1884. One can imagine the amount of labor, both animal and human, necessary to build the structure. Wood for the framing and interior had to be cut, the heavy stones quarried, then hauled, perhaps miles, to the church site before parishioners constructed the building. There must have been a joyous celebration in the community at the time of the completion of the building.

More than anything, I remember the other young people who attended the church. There was my sister, Pat, one year younger than myself, and Dott, my older sister, and her boyfriend, Fat, and his twin brother, Skinny. They were both on the high school basketball team, which made them instant heroes in my eyes. Then, there was Fat and Skinny's sister, Dolley Mae, who was Pat's age. Dolley Mae was an excellent pianist and often played at services or gospel sings at the church. She could, with ease, play any song the congregation

Berwick Community Church

selected from the hymnal. Her ability still impresses me. Then, there was Denny, who now owns the farm next to mine, and Naoma, and Norma Jean, and Ralph and Leon and Jean and finally, Gary, who now bales the hay on my farm.

Once we began attending the church, we rarely missed Sunday morning services. Worship services were normally held twice a month, Sunday School every Sunday. The pastor, a gentle spoken man, lived in a nearby town, where he was employed as a school administrator. He was a man chosen by God. He knew his scripture well and the way to salvation. He was unbending as he instructed and exhorted us to live an irreproachable life.

On Sunday mornings neighbors from throughout the community would collect at the church, each family arriving separately in their family cars and pickup trucks. The women and children would immediately climb the few concrete steps to the the sanctuary while the men lingered outside greeting all who entered, while also discussing farming, the weather, crops, and having their last puffs on cigarettes. Everyone who entered the church had to run their gauntlet of smoke.

First came Sunday School. All in attendance gathered upstairs for an opening prayer, a hymn, and devotion led by the Sunday school superintendent, who would then announce that classes would begin. The young people always met downstairs, the youngest children to the back of the basement, we older kids toward the front. Our instruction came from reading our Bibles, our Sunday School manuals, and listening to the teacher.

Sunday worship service followed Sunday School. The leaders of the church prayed. The congregation sang hymns. The preacher preached. The lost were saved, and all of God's children received inspiration for having been in the House of the Lord. What a blessing. No week in our household would have been complete without attendance at church.

Often, on Sunday evenings, we young people returned to church where one of the church leaders, Erma, led a youth group called Christian Endeavor. (I can remember we attended this Sunday evening class long before we began attending morning services.) At Christian Endeavor we learned and recited Bible verses, read the scriptures, and Erma talked to us about the necessity of us being "good." But Christian Endeavor was also a fun time for we got to talk and play games, as well as learn about religion.

All was not Bible study or worship. Everyone in the community loved to eat, and church functions provided another opportunity. It seems, even in retrospect, that there had to be no special occasions for an after-church gathering. People just brought plates and baskets full of food which the women placed on the long tables in the basement. Roasts and hams and chicken and vegetables, almost all of it grown on local farms, filled the tables. When it was time to eat, the minister or one of the church elders would offer a long and thankful prayer, and then the men would pass along the long tables filling their plates to overflowing, followed by the young boys, the girls, and then the women.

After eating, an afternoon songfest would often erupt in the sanctuary. Members would gather in the sanctuary and would call out their favorite songs for the congregation to sing, and Dolly would play, sometimes for hours, the hymns requested by the congregation. Mixed in the song service would be solos, duets, even quartets, some of them impromptu arranged, but always singing old-time favorite hymns to the praise of God. Others gave testimonials about the wonderful things God had done for them.

Berwick Community Church

But these afternoon services, although dedicated to the glory of God, would sometimes become boring and at times the boys, discreetly, of course, would excuse themselves and amble toward the outhouse that was across the road from the church. The outhouse was, for the most part, just a smelly mess, a two-seater on each side, making a total of a four-holer. It was normally used only by those with emergency needs or on those occasions when church attendance took a long time, such as Sunday morning services followed by dinner on the ground followed by a songfest. It also served as a meeting place for young boys—boys soon to become men—to talk about girls and tell dirty jokes and teach each other bawdy songs.

On other occasions when we kids had little to do at church and were feeling especially brave, we would wander out into the cemetery past the old huge white oak tree that stood guard at the gate. We always felt eerie when we did this, wondering who all those people were underneath the headstones, with dates inscribed clear back into the previous century.

The highlight of the year was Christmas, and the church always sponsored a Christmas play. The play reenacted the birth of Christ. Practice for the play began several weeks before Christmas with all the young people gathering on week nights to learn our parts and walk through the pageant. I was a shepherd in several of the presentations and had to wear a long bathrobe with a turban wrapped around my head. All of us seemed always to be hesitant and nervous about our participation, but proud of our accomplishment once the pageant ended. On the night of the pageant, the church overflowed with a most attentive and appreciative audience. And the evening was always climaxed with the arrival of Santa Claus who gave out large bags of apples, oranges, and candy to the kids.

Return to the Farm

I was a preacher at the church for a while when I was in college. I had received the "call" to become a minister while I was in college and, during a string of a few months, while the church was without a minister, I went back to the church on weekends to preach. Members of the congregation, friends all, were exceptionally kind and doting. At the end of the service I would stand at the back door and shake hands with everyone as they left. "That was a good sermon," they would say, or "You'll really be a good minister," or "It's wonderful to have a local boy who can preach so well." Even my grandmother, who was a staunch member of her own congregation back in Oklahoma, came to hear me preach one Sunday, and told me that I would never have another church that would like me so well.

In fact, I intended to be a minister and studied for the ministry while in college. However, I eventually discovered that I had made a poor choice about the seminary I should attend and thoughts of another vocation and other worldly endeavors entered my mind, and so I changed the course of my life—to a non-church vocation. However, little did I realize the underpinning that this experience would give me. Now, perhaps, years later, the circle of life may be closing.

Although over the years I've visited the Berwick church irregularly, more often when my parents were alive, less often now that they are not, the church has continued to change and adapt to the changing community. For a period of time attendance at the church dwindled, and the church was in danger of closing permanently. A new minister, however, revived the church and attendance soared.

In the early 1980s the Cumberland Presbyterian hierarchy went through a reorganization and disassociated itself with several of its smaller, rural churches. The Berwick Cumberland Presbyterian Church was one of the churches from which it severed its association, so it is no longer a

Berwick Community Church

Cumberland Presbyterian Church—now it is just the Berwick Community Church.

More recently a disagreement ripped through the congregation and some members left, starting another church not too far away. But, over a few years, both churches grew so that now there are two thriving congregations where before only one church served the community.

The Berwick Community Church still serves the needs of the community. Regular services are held with a full-time pastor; a youth ministry has been inaugurated, and every year a few burials and an occasional marriage is held in the church.

The biggest change for me, though, is in the cemetery. The cemetery has grown by two, three, or four people a year. I can remember the faces, beckoning smiles, and peculiarities of most of the people who are buried in it. Even one of my high school classmates is buried in the cemetery. Near the front gate the massive old white oak tree is losing its vitality, for it has been hit by lightning; every winter a few more of its limbs drop off, and in the springtime fewer green leaves bud forth. But the change that rivets my attention is that fact that now both of my parents lie in the cemetery beside the church. When I drive past the church, on the way to the farm, I am always reminded of them.

Clear Creek

I saw Sid Hill, a trapper, trudging up Clear Creek the other day, two small lifeless muskrats dangling from his left hand. A few minutes later, when I peered into the bed of his old beat-up pickup truck, I discovered another animal, a brown, soft-textured, medium-sized beaver he had trapped earlier in the morning. Even dead, the animal was beautiful. Seeing the animals reminded me of the winter many years ago when I trapped a few muskrats out of the creek. I was proud, then, that I had mastered this elementary trapping feat. Muskrats are the easiest of creatures to trap, with mink and beaver much more difficult. Sid, having trapped for many years, is equally adept at trapping any animal.

The price paid for raw furs, Sid told me, is negligible these days, and so he traps only for the pleasure of it and to rid the creek of pests when an overabundance of fur-bearing animals appears. I was a little annoyed because he trapped a beaver. I enjoy looking for fresh beaver signs on my property, for the small saplings they have gnawed down or larger trees they have "rung" with their teeth, turning the bark of the trees into food. Yet, I know they can be destructive. Several years ago too many beavers in the creek destroyed almost every young tree along the creek. Years passed before the forest floor rejuvenated itself. But a few animals need to remain though and Sid seems to know what that number is because he's been trapping Clear Creek for years, and signs of muskrat and mink, raccoon, and beaver still remain along the creek bank.

Childhood memories of adventures along Clear Creek still linger in my mind. Normally, except after torrential rains that turned the stream into a raging torrent, the Clear Creek was a wonderful place for fun and excitement. As kids we would swim in chest-high holes on hot afternoons to cool off

Clear Creek

from the sweltering heat and the high humidity. We would splash the cold water on each other, search for crawdads underneath the rocks and pebbles on the bottom, and occasionally feel the nibble of small minnows on our legs. Occasionally too, on a Sunday afternoon outing, I caught a few small mouth bass from the riffles and even one winter, as mentioned, trapped a few muskrats from the creek. Neighbors still gather by the creek occasionally on summer evenings to picnic by lighted bonfires.

I was alarmed that Sid was wading through one of our old swimming holes in knee-high rubber boots. Was my mind playing tricks on me? Was not that "hole" much deeper when I was a kid? Did not more water flow down through Clear Creek forty years ago? And, documenting an even earlier time, those old-time neighbors, who used to gather at our house for dinner after church when I was a kid, talked about running and jumping from the bank into holes in Clear Creek when they were kids and never touching the bottom. Now, Sid was wading in these same pools with boots that were only knee high.

Based on what Sid and I could determine, the flow of Clear Creek was becoming less and less. Would the creek continue to exist, we asked ourselves, as we stood beside his pickup truck and talked about the creek. Would, in a few years, Clear Creek become a dry creek?

To learn more I contacted a water quality specialist with the Department of Natural Resources in Jefferson City. This public servant, who obviously had monitored Ozark streams for many years and felt a keen affinity for the quality of water, added to my knowledge of streams. This official stated that the level of water in a stream is influenced primarily by the amount of rainfall in a given area. The Ozarks, with an average of roughly forty inches of rainfall yearly, has sustained

Return to the Farm

that amount for as long as records have been kept. So the amount of water that flows down Clear Creek, he continued, is probably about the same as when I was a boy. But water levels can fluctuate, he indicated, based on the use of land along the creek. When forests cover the hillsides and valleys beside a creek, the rainfall that reaches a creek flows slowly, being first absorbed into the forest floor and then slowly released into the stream. But when the forests and pasture are gone and nothing remains to hold rainwater in the soil, the water, after a rainstorm, immediately runs into the stream. And, if the rain is heavy, erosion takes place. The streambed and "holes" are filled with the soil, gravel, and rocks carried by the turbulent run-off and flood waters. Sometimes, he stated, hundreds of years elapse before a stream can wash away the gravel and rocks that accumulate after a series of heavy rains. This, he speculated, is what is happening to Clear Creek.

"But what about the quality of the water that flows down Clear Creek," I asked. I repeated to him what one of our old-time, lifelong community members, Basil Ferguson, had told me: that on occasion he had seen foamy pollutants, toilet paper, and human discharge being carried down Clear Creek. "Are there still fish and minnows and crawdads in Clear Creek," I asked? (Undoubtedly, I've been negligent in not taking my fishing pole down to Clear Creek and finding out for myself what fish still lurk in the riffles and pools of the creek.) Though the agent was not aware of any "fish counts" on Clear Creek, he did send me the results of a survey of a nearby Ozark stream that indicated roughly fifty species of fish and aquatic life existed in that stream, a number that was down by nearly twenty species from a count taken twenty years earlier. Aquatic life, he indicated, is a measure of water quality. The more and varied the fish and insects that live in the water, the higher the quality of the water.

Clear Creek

A few days later another worker at the Department of Natural Resources explained to me even more about the overall quality of water in Ozark streams. This official helps coordinate a volunteer program called "Steam Teams," a successful effort to monitor stream conditions in Missouri with the aid of, and as a way to educate, young people about environmental issues related to water quality. This lady told me that all kinds of critters—worms and insects, snails and clams (she called them crustacean and micro-invertebrates)—live in the stream beds, buried in the sand and gravel under the flowing water. Each section of the creek, from fast running current in riffles to slow moving water in deep pools, has a different kind of micro habitat, and the inhabitants vary just as do the different animals that dwell in the deep woods, open prairies, and deserts. But most water creatures, she continued, are small, under an inch in length and with interesting names: helgramites, mayflies, water pennies, caddis flies, black flies, scuds, snails, clams, and worms. I was surprised at what she said. A plethora of insects of which I was largely unaware could exist in Clear Creek.

Because these aquatic creatures are sensitive to pollution, a survey of the creatures would serve as a guide to the water quality in Clear Creek. A pristine stream would have varied species in abundant numbers; a polluted stream less variety and numbers. Her explanation made sense to me.

This official even sent me specific instructions on how to conduct a scientific survey of the creek, complete with a list and photographs of the little creatures that were known to exist in Missouri streambeds. Our task would be to survey for the little critters by collecting our specimens from three different areas/habitats of the creek since the creatures in a creekbed could vary, even in a short distance, depending on the flow of the water. The fast running water in riffles carried the most oxygen, and thus contained the most sensitive or least pollution-tolerant organisms. Steady running water at the head of a riffle contained somewhat less oxygen, but more specimens than in the riffles; while slow running water in shallow pools contained the least oxygen, but probably the most, although least hardy, species. Her instructions for performing the survey were complete. We were to select three different sites, and, as a way to dislodge the little critters, stomp and kick the water, gravel, and sediment in the streambed in a ritual the instructions called a "scientific" dance, hoping to catch the creatures in a net, and identify them, all with an eye toward determining the water quality in Clear Creek.

Karlene and I decided we would conduct the water quality survey ourselves. One sun-brightened morning, the first Saturday in October, on a day, unfortunately, when no grandchild was available to help, we plopped out of bed, more than a little self-conscious about what we were about to undertake and apprehensive about any anticipated results. We, nevertheless, agreed to pursue our task. "It's a scientific endeavor," we told each other.

We gathered up our equipment—the seine and magnifying glass supplied by the Department of Natural Resources, our booklets, notes, clear plastic milk jugs, rubber boots, faded blue jeans—and headed for the farm. On the way we decided that Karlene would take charge of the seine, making sure that none of the little creatures lurking in the bed of Clear Creek would escape from our project while I would stand upstream and perform the environmental dance, tromping on the bottom and kicking loose the rocks and pebbles and dislodging any aquatic life that might be hidden underneath.

The only tasks that remained after arriving at the creek were to cut down a couple of five-foot saplings to serve as handles for the kick-seine and don our wide-brimmed hats.

Clear Creek

Karlene pulled her knee-high wading boots on while I laced up an old pair of canvas shoes. Soon, we were trudging up the creek. Anyone who saw us would certainly believe we were the most eccentric couple in the Ozarks! Still, we were determined.

Guided by our instructions, we sought out three sampling sites. All, we quickly noted, would be within 300 feet of the bridge. We waded no more than thirty feet upstream in the slow running, shallow water before declaring the spot below our feet the first area for our investigation. Carefully, Karlene unrolled, then tilted her three-foot kick-seine backward to form a pocket so she would be able to catch all the aquatic critters that might float downstream. Meanwhile, I anchored the bottom of the kick-seine to the streambed with rocks, and then began our experiment by kicking a three-foot by three-foot area of the streambed immediately in front of the seine, loosening soil and rocks to a depth of six inches. The more vigorously I kicked, or "danced" as the instructions called it, the more the streambed boiled with brown sediment, small rocks, and muddy water that floated downstream into the seine Karlene held. Would we catch anything in the seine, I wondered, as I "danced" this environmental investigation in the streambed at the base of my farm?

What a revelation occurred as we ended our dance. In a moment the dirty water passed us by and we could see, amid the rocks and debris in the net, a variety of small, hitherto unforeseen by us, little water creatures couched in the pocket of the net. A world of creatures lived below water level in the streambed of Clear Creek, much as they must have existed eons ago. Undoubtedly their forebears had existed long before man and might exist long after mankind becomes extinct. We identified the crayfish and snails, but had little knowledge of the other little creatures, ranging in length of one-fourth to two

Return to the Farm

inches. Carefully, we placed a dozen or so of the creatures into one of the plastic jugs. We would identify them later.

Amazed with our initial find, we waded further upstream, easily walking through the very hole of water I had swum in as a kid. This second site put us at the head of a riffle. We repeated our procedure as at the first site. Karlene placed the kick seine in the water and tilted it back while I anchored the bottom with a couple of stones, then began kicking and tromping the swirling water until I had cleared a three foot by three foot area six inches deep in the streambed. Then, carefully, when the water had cleared, we separated the pebbles and debris in the net from the tiny creatures lodged in the pocket. We added to our collection by placing a dozen or so little creatures in our second milk jug. More success. Finally, we moved into the sparkling, fast-flowing riffle itself. Within a few minutes our "kicking and netting" had garnered us a collection of another dozen or so water creatures.

Astounded by our apparent success, we carefully waded back down the creek and headed for an only partially-decayed log on the bank, where we rested before examining our bounty, keying it to the cards and photographs from the Department of Natural Resources, and identifying our treasures.

Nor was the keying easy. We quickly released the crayfish and snails back into the water. We knew what they were. The aquatic worms, although much slimmer than the earthworms found in the soil, were also easy to identify. I just didn't realize that worms also lived in streambeds. The sowbug and riffle beetle, too, were easily keyed to their picture. They resembled insects I had seen on decayed trees. These creatures, however, spent their whole life in streambeds, under water. But the damselfly, scud, stonefly and caddis fly and dobsonfly, while easy to distinguish from each other, were not easy to

Clear Creek

identify. All were from a quarter of an inch to two inches in length, some juveniles, some adults, and all with various tails, legs, platelets, and antennae. Still, we were overjoyed to have found a greater variety of crustacean and invertebrates than we had anticipated. Some of them undoubtedly were in the very sensitive group, indicating the water quality of Clear Creek was very good. We were amazed with our discoveries.

But the discovery that the water quality of Clear Creek is very good only partially alleviated my concern about the creek. Our water samples had come from the very same area, the very same swimming hole, where I had laughed and splashed and swum as a kid. Now, Karlene and I, as had Sid earlier, trekked across the pool. The water was not even knee high to me. No water overlapped into Karlene's knee-high boots. Clearly the water flowing down Clear Creek, albeit in this drought year, was less than in previous years. I concluded, Clear Creek is drying up.

Ironically, another water problem beset our farm this same year. Because of the drought and our enlarged herd of beef cattle, the pond threatened to go dry. It was necessary to drill a well, have the electric co-op install electricity and a meter loop on the farm, and provide a water tank to the cows. "You'll have to go 500 feet for water," the well driller told me. "Used to be you only needed to go 170 feet in this area, but that aquifer has about dried up. The farms with old wells are having to redrill now, going much deeper," he said. He told me about a neighbor, one who raised mega-numbers of chickens, who faced a near calamity that summer when he suddenly discovered one day that his well had gone dry. "But I can get you a steady stream of water at 500 feet," he repeated. 'The water level has lowered."

He did have to drill deeply. I was at the farm observing when the driller, after two days of work, hit an aquifer of water

123

that at first trickled, flowed, and then gushed with water. "That's at 478 feet," the driller said. "You'll have excellent water from now on," he added, satisfied that he had hit a gusher at the depth he had predicted. I would no longer need to worry about my pond going dry in summer. I now had a good well on my farm. My neighbor, Dennis, who raises acres of tomatoes and other vegetables, as well as gourds, and who irrigates his crops, showed an active interest in the drilling of the well. As soon as we hit water, he too requested that a well be drilled on his farm, at the depth of the new aquifer. I can't help but wonder how many other people, over the years, will be drilling into this aquifer. How long will it be before the water in this new aquifer, too, is depleted?

But the inescapable fact exists, I think, that my new well, coupled with all the other wells along Clear Creek Valley, are forcing the underground water levels lower, robbing the springs that dot the countryside of their source of water, and thereby diminishing the flow of water in Clear Creek. In fact, it is probable that some of the water that makes its way into Clear Creek actually filters down through the creek bed into the ground, actually diminishing the creek as it flows. There is no doubt in my mind that in fifty years Clear Creek will cease to flow. In fifty years it may serve as a dry creek, a bed for rushing water after a heavy rainstorm, or as a drainage ditch for other unwanted water. But I doubt that it will exist as a pristine, crystal flowing stream. Other uses for the water, in agriculture, industry, or for golf courses will pull the water from the ground before it reaches the stream, and divert the water to other uses. In fifty years Clear Creek will be a memory, with old men telling their grandchildren about the time they swam, hunted, fished, and trapped in Clear Creek and families picnicked on its banks.

Clear Creek

My anxiety about Clear Creek increases and I feel helpless. I feel I'm a part of the destruction of the creek and am unhappy with this deterioration. I see no adequate measures to reverse the demise of Clear Creek. This fact gnaws at me.

But, there is more to understand about Clear Creek!

I have become intrigued, perhaps mystified is an even better term, about the geology of Clear Creek and what it represents. For, I have recently become aware of how Clear Creek came to be and am awed at the vast eons of time necessary for its formation. For the consequences of the geology, I have learned, affect not only the creek today, but the total environment of the Ozarks.

For me, the inquiry into the geological formations began when I was a boy and would find small fossils of shells embedded in limestone rock that littered our farm. How, as a kid, I wondered, could seashells become a part of a limestone outcropping on our farm? It would be years before I knew the answers.

My geological friends supplied the explanation. The Ozarks, a region that stretches from the Missouri River Valley in Missouri to the Arkansas River in Arkansas and from Illinois to Oklahoma, is, geologically speaking, dominated by a Karst topography. Karst topography is a landscape both rugged and fragile: rugged because much of the Ozarks is dominated by scenic craggy hills and deep valleys, and fragile due to fissures that allow rainwater to pass, sometimes with almost no filtration, directly into the aquifers below—but I get ahead of myself.

Geological theory holds that the bedrock underneath the Ozarks is at least 3.5 billion years old and was formed slightly south of the equator. Since that time, the North American continent or crust, as geologists call it, has shifted continuously ever so slightly northward, plodding relentlessly, almost imperceptibly except with a measurement that can only be ascertained in millions of years—perhaps the continent moves only a few centimeters each year. For those of us used to measuring distance in miles per hour, the shift is imperceptibly slow. And, the whole continent continues its northwest course to this day.

The seashell fossils are a testament to that fact that while making its northward trek, the Ozarks were, more than once, a seabed covered by water. Time and again the seabed raised upward to form a land mass, then sank again to form another sea. When the Ozarks was a seabed, a variety of calciferous seashells formed, lived, and died in the seas and floated to the ocean floors where, under compression of later sediments, they became the limestone formations so prevalent in the Ozarks. But, a few of the seashells did not turn into limestone. These few fossilized, and thus today we find evidence of these small sea creatures, fossils, embedded in limestone ledges throughout the Ozarks. I can search the bluffs above Clear Creek or along other creek beds or highway right-of-ways and find fossils of seashells hundreds of millions of years old.

This last uplift of the sea, geologists believe, occurred about 300 million years ago, forming a dome in the center of what we now call the Ozarks. But, this uplift also developed fissures, and abrupt fractures. Over millions of years the chemical action of rainwater, made slightly acid when contacting the organic matter on the surface of the earth, has eaten away at the limestone, forming creeks and valleys and leaving beside them hills and plateaus. Underground, as rainwater passed through the open fissures, the acid water ate away, ever so minutely, at the limestone. Over time, eons of time, the dissolving of the limestone by the acid rain formed pockets, or sinkholes in the earth, which became caves, natural arches and, finally, in the valleys, streams. Thus, the simple action of acid rainwater formed much of the rugged beauty of the Ozarks.

Clear Creek

Such action continues to shape Clear Creek and the Ozark region.

This geological action never ceases. Just as water continues to flow through Clear Creek, so also rainwater still runs through open fissures to underground reservoirs. This is the water which Ozark residents tap into for drinking, household use, and watering livestock. When this water becomes contaminated on the surface by chemicals, rubbish, trash, and leaky septic tanks, little cleansing occurs by percolation through the soil. Drinking water becomes polluted. People and animals become sick. The earth deteriorates.

Thus, when I go down to Clear Creek, much enters my senses. Clear Creek has given me much. In the last few million years Clear Creek has meandered over much of my farm, laying down rich soil, erosion from upstream. For the excellent soil on my farm, I have Clear Creek to thank.

Then, there is the beauty of the creek, which, for me, still holds an almost romantic attachment. The serenity of the area and the beauty of the creek touches through to my soul. The rippling waters, at any time of the year, continue to delight my senses. In the spring the blossoms of the redbud and dogwood trees that line the creek herald the coming of a new year. In the summer the cold water offers a retreat from the stifling heat of a muggy day, and I can still skip a rock on the surface of the creek and be refreshed by wading in the water even while overhead I hear a squirrel barking. In the fall the leaves of the overhanging trees turn autumn red, golden yellow, or blazing orange; black walnuts fall to the ground along the creek, and I must compete with the squirrels to get a taste of the distinctive nut. In winter, the water is at its clearest and I can look down onto the rippling surface and see the reflection of a blue cloudless sky—broken by the mirrored image of one of the towering white-barked sycamore trees that line the bank.

Return to the Farm

As I peer into the water, a crispy leaf may float down from the sycamore tree and settle on the stream surface, to be carried away by the flowing water.

But, more than anything, it is the time element that mystifies me. For, there is a sense, a part of belonging to planet earth, that cannot be understood by the brain. How long is a million years, a billion years, or 4.5 billion, the time scientists believe that planet earth has existed? (Cosmic timeframes are even longer.) I do not doubt that the earth has existed that long. I just cannot fathom that length of time. From a geological time frame, say a few hundred thousand years, it makes little difference what mankind does to the creek or the environment. Nature will, in the end, have its way. But, my time frame is shorter. I can think in concrete terms only about my generation, and that of my children and perhaps a generation or two after that—then my ability to perceive time collapses. But I am happy to look that small distance into the future. Indeed, I feel that that we all must. For, I am a part of the future and through my actions, the future becomes a part of me. And, I am disturbed now about some of the problems we are causing in the environment and with farmland, for if we do not change, not only will my life be paled, but the lives of those who follow after me will necessarily be diminished.

Is it any wonder that when I am at the farm, I often take time to visit Clear Creek?

Clear Creek

Farm Today (2004)

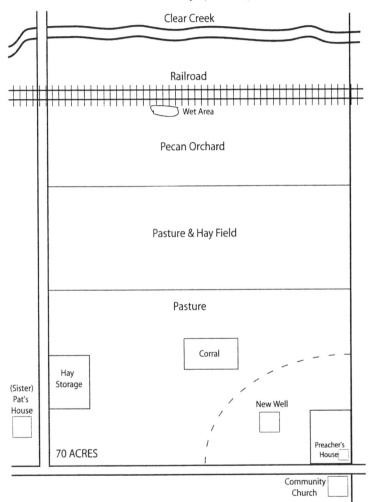

Who Owns My Farm?

I own this farm. It was an inheritance, a gift, given to me by my parents who had owned it and worked it and treated the land in a special way that only a hard-working farm couple, in love with their craft, can. When my parents died, in keeping with their wishes, the family farm was broken into three as nearly equal parts as possible, with one part going to each of us three children.

My father, like so many other farmers, carried an extraordinary commitment to his farm life. He chose, with some hardship to his family, the vocation and spent the rest of his life managing his farming operation. He, like many other farmers, was particular about his land, the product he produced, and the quality of food he raised. Looking back, although he was joined to the land through legal ownership, his attachment extended far beyond an economic and perhaps even an emotional tie. Farming was an integral part of his being. Although legal owners of the land, he and other farmers felt a responsibility beyond that of their own tenure—as if the vocation itself contained a mission they must perform. Dad loved his work and was totally dedicated to farming. He expressed concern about passing his farm on to others who would appreciate the land. What else could account for the commitment, long work hours, and small economic return? Long ago I decided that many farmers exhibited a stewardship for their profession that extends well beyond that of land ownership.

I felt an inkling of this dedication when I inherited the farm. When my older sister read the will my parents had made, at the funeral home shortly after my mother's death, there welled up in me an unexpected pride, a passion that I had not expected about owning and working on a farm. I decided, instantly, to carry on, albeit in a small way, the family's tradition of farming.

Who Owns My Farm?

But, as a new farm owner, I became intrigued with the concept of land ownership. I wanted to know about the previous owners of my farm and about the traditions of farming that I had promised myself I would uphold.

I knew that the first Indians had lived in the Ozarks at least 10,000 years ago; in all probability, some of them occasionally traversed the farm, especially along Clear Creek at the end of the farm. But for unknown reasons this tribe of Indians disappeared, leaving traces of their existence only in caves and overhanging ledges above streams. Still, from that time onward, other tribes inhabited the Ozarks and perhaps, occasionally, lived on my farm.

We children used to comb the newly plowed fields below the farmhouse and find arrowheads in the soil around the spring. These arrowheads, we now believe, were discarded by Osage Indians who very likely resided in a seasonal hunting village established near the spring.

Before they were pushed from the area following the Louisiana Purchase, the Osage had called the Ozarks home for over 700 years. Before arriving in the Ozarks, their tribal traditions state, they had resided in the stars with Wah-Kon-Tah who recognized that the earth needed care and an overseer, a tribe of people to be stewards of the land. So Wah-Kon-Tah moved the Osage from their home in the stars to the Ozarks, a land the Osage called the Middle Waters because of the many rivers, where they could become custodians of the land. In the Middle Waters the Osage were to look after the buffalo, the elk, and the deer; to become brothers to the bear; to be thankful to the fish; and to eat the squash and maize that they cultivated. They were to take no more from the land than they needed. When they hunted the buffalo, they divided the meat among the members of the tribe: the tongue, the hump, the stomach, and the fore and aft quarters. All food was distributed among

Return to the Farm

tribal members according to need so that in times of plenty all had food to eat and in times of famine, all hungered together. It was a communal system with "the little old men" of the tribe, as the wise ones were called, giving thanks to Wah-Kon-Tah for all that he did: for aligning the forces of nature so that the tribe could continue to eat, for life and protection after death, and for help against the cycle of the seasons, especially when the cold winds and winter weather bore down on the Osage. For 700 years, the Osage prospered as a tribe in the Ozarks, some members residing at least occasionally near a spring and on my property.

But the Osage did not own the land. They simply possessed it, controlled it, and used it according to their needs. They could even fight over the land with other tribes. But they did not own it. They could only be stewards of the land, passing it on to others who would follow. They found the concept of land ownership that the first white intruders eventually imposed on the land a foreign notion.

The first contact between the Osage and the white man occurred in the early 1700s when French traders arrived from St. Louis seeking furs and pelts. The Osage and French maintained an amicable relationship for nearly 100 years, until such time as the United States government completed the Louisiana purchase from the French in 1803 and the land inhabited by the Osage became a part of the United States. Immediately after the purchase, President Thomas Jefferson spearheaded a movement to settle the Indian problems—but Thomas Jefferson recognized only "Indian problems" east of the Mississippi River. In 1808 to pave the way for the transfer of eastern tribes into the Ozarks, the government forced a treaty with the Osage, taking away much of their land and forcing them first onto a reservation near what would become Kansas City and later, Oklahoma. This allowed the eastern tribes to move into the Ozarks. This shuffling of tribes, of course, lasted a few years until the new tribes,

Who Owns My Farm?

the Kickapoos, Delawares, and Shawnees, were also pushed westward.

So our family moved to the farm when European or western land use and concepts prevailed. Under western land use, owners have a legal right under the law to do much as they wish with land. They can buy it, sell it, rent it out, raise families on it, raise food which they either use themselves or, as on most farms, sell in exchange for money. They may do much good with the land by raising indigenous crops, building the fertility of the soil as they farm, and controlling the water supply so that it is not slowly depleted. Or, they can abuse the land. They can put too much seed on the land and deplete the soil, pour too much fertilizer on the land until it becomes hardened, spray it with chemicals and kill the weeds and organic plant life, and strip the land of trees and shrubs and displace the wild birds and animals. And, although now less common, farm owners have been known to poison the streams and kill water life and fish. And, at death, in accordance with inheritance laws, landowners may pass the parcel on to another person.

In keeping with the traditions of land ownership practices handed through our European heritage, I, too, am concerned about who will get this parcel of ground after I can no longer use it. For one thing, my wife is now the "joint tenant" of the land. Several years ago, she wanted to become part owner and so we set this down in a legal decree that she filed at the courthouse in Neosho. If I die before she does, she becomes the owner of the farm. But we also agreed at the time, with verbal promises, that she would pass the farm on to my children, with her three children eventually gaining possession of the house in which we reside off the farm.

My father and mother purchased the farm from an old maid schoolteacher named Addie Lou Montgomery, the last in

Return to the Farm

a line of Montgomerys who had owned the land. The Montgomerys, who were among the first settlers in Clear Creek Valley, arrived in the 1840s and retained possession of the land until they sold it to my parents. I can barely remember seeing the woman, a small, frail and wrinkled old lady, who was reluctant to sell her land, partially because the Montgomery family cemetery was located on the farm. The cemetery, even then already neglected with toppled and broken headstones and overgrown with brush and trees, was, nevertheless, the resting place of many of her forebears. She told us at the time that the cemetery contained over 125 burial plots, but often only wooden crosses and field stones had been placed as grave markers; we never found over a dozen or so headstones, the earliest one dating from 1845. But Addie Lou was very particular about who would eventually own her farm. Since Dad was a sincere family man with a good wife and children who needed the out-of-doors, he passed her test, and the land changed hands, not just from one individual to another, but from one family to the next.

In 1954 the land was legally transferred from Addie Lou to my mother and father who lived on the farm for over 40 years. Thus, this piece of ground has had very few owners. Now, the land is my responsibility and it is redundant to say that caring for the land takes time and energy and that the economic returns are meager. Yet, I remain acutely aware of the paucity of return even as I continue my effort to maintain and increase the productivity of the farm. Something more than ownership drives me to farm.

It is also true that, in terms of production, the produce sent from this farm into the agricultural marketplace is much less than it was when my Dad farmed this piece of property. Every year I ship a few feeder calves to market and occasionally sell a few bales of hay. Compared with the massive

Who Owns My Farm?

amounts of wheat, corn, and soybeans that are produced in the great Midwest breadbasket and the number of feeder calves sent to market throughout the United States, my commodities are infinitesimally small, counting for very little, except to me. Still, I do send the cattle and hay and hopefully, eventually, pecans to market. And a distinct feeling of satisfaction wells up in me each time I send produce to market. The toil of my labor supports not only me, but also contributes to the welfare of others.

I find my contribution paled even further when I realize how my farm is now tied so directly into the world economy. Local agricultural news reporters give weather, planting, and even political conditions in countries around the world before reporting the prices of commodities as established in Chicago. The prices for barley, rice, soybeans, and corn in the United States, for example, are not set without ascertaining the outlook in nations around the globe. The same holds true for the futures reports on beef, the report that most holds my attention. Prices in the United States are not established without first determining trends in Australia and New Zealand, South America, and Europe. All is interconnected. The vastness, complexity, and interrelationships of markets dwarf my efforts even further.

Still, I fight to maintain the belief that the produce from my farm is more than an economic enterprise, more than a minuscule cog in a great, worldwide, agricultural production system. I want to believe that, through my efforts, this farmland contributed significantly to the the health, welfare, and nutrition of others, and that my efforts, in some small way, helped. But this is where I become fearful. For, many farmers, and I am one of them, believe that the world's agricultural system is running amuck.

Return to the Farm

I've read and believe (without much documentation, to be sure) that there are fewer people in the United States setting agricultural policy than there were in the old Soviet Union. A few university professors, members of Congress on agricultural committees, and the heads of a handful of multinational corporations can decide what commodities will receive subsidies and thus become profitable, or what agricultural research will be conducted, and the decisions of these few establishes the agenda for all of the agricultural community.

I know many believe that with all the worldwide interconnections and bioscience and food manipulation, a great new day is dawning in food production. They claim that food availability will rise substantially at less cost.

But why, I wonder, is there still hunger in the world? Why are not the hungry being fed from this bountiful agricultural promise that is so apparent? We were downtown in a large American city the other day, and my daughter showed me where, under a fenced underpass, many of the homeless used to congregate and live. But city officials decided to expel these tenants and so we saw homeless men, women, and children dispersed throughout the downtown area of the city. Even in the Ozarks, churches and agencies exist that are dedicated totally to feeding the hungry children, adults, and the elderly. A 2001 report from our own Department of Agriculture found that there are 8.5 million people suffering from actual hunger in the United States, with many, many more millions plagued with an insufficient amount of food. Many of these were children. How can a land that has so much bounty speak of successful agricultural programs when so many people are hungry? The question perplexes me.

Right after the 2000 presidential election, I saw George McGovern on one of the morning TV news shows. He was on

Who Owns My Farm?

the program only a few moments, but was touting an agricultural assistance program designed to feed hungry school children throughout the world. The details of the program were vague, but our new president had just requested that George McGovern work with this program. My interest was pricked both by the image of the man and the program he was championing—an initiative to ensure that every school child in the world would receive a healthy breakfast. I thought this, if practical, a very good idea.

This was the first time I had seen or heard of George McGovern for years. The sight of the man brought back many memories from the era of our nation's political turmoil.

For those who may not know, George McGovern, a distinguished veteran, was the presidential candidate for the Democratic party during the Vietnam War, running against Richard Nixon. George McGovern fought against Vietnam policies. He, of course, did not receive many votes, some say, because he didn't "act" very presidential during the campaign. At any rate, he lost the election, and then, in his next senatorial bid to retain his senate seat, was defeated again.

George McGovern ceased to make press coverage after that time, except for a brief period when his daughter, an incorrigible alcoholic, froze to death outside a bar on a bitterly cold night in Minneapolis in December 1994. The death brought great sorrow to George McGovern and his family, and the ex-senator began to speak out, compassionately, in memory of his daughter, on the devastating affects of alcoholism on individuals, families, and the nation.

But, I would learn, it was his efforts for alleviating world hunger that had kept George McGovern occupied for several years and it was in this capacity that he was on the morning news show.

Return to the Farm

I decided to contact George McGovern's office to learn more about his proposed program. Initially, I thought I could find him by searching the Internet, but, although I could find out a lot about him and read excerpts from several of his speeches, I could not find his current address.

Even enlisting the help of local representatives of our congressional delegation resulted in a fruitless search. A young man in one office confessed he had never heard of George McGovern; another, after a three-day response time, informed me that George McGovern worked at the State Department and even gave me his telephone number which, when I rang it, was answered by a surprised State Department official who acknowledged that he had heard of George McGovern, but was perplexed as to why I had called his telephone number. However, he must have had an inkling of where McGovern might be, for he conjectured, "He might be at the United Nations office in Rome," before wishing me the best of luck in finding George McGovern.

A second attempt at searching the Internet eventually turned up an office address of the World Food Program, and I came to believe through the information on the Web that George McGovern was an appointed ambassador for the United States to that organization. A phone call to the office in Washington confirmed what the Internet indicated.

A volunteer at the Washington D C office of the World Food Program, with offices in New York City and Rome, told me that while in the Senate, George McGovern and Senator Robert Dole had co-sponsored legislation establishing a worldwide food relief agency. For years, this agency had worked in conjunction with the United Nations, but was now an independent organization, the World Food Program, working with governments throughout the world to eradicate hunger and

Who Owns My Farm?

poverty George McGovern, while not in the national spotlight, had spent many years supporting the organizations and efforts to feed the hungry throughout the world.

I received an educational kit from the World Food Program about hunger in the world, the effects of hunger, especially on children, and current efforts to eliminate hunger. The World Food Program estimates that today an appalling number of people, 800 million worldwide, are affected by hunger—hunger brought about by disease, drought, civil unrest, and war. The effect on victims, particularly children, is horrendous, with thousands throughout the world dying daily. Meanwhile current efforts to eliminate hunger, although encouraging, remain dismally inadequate.

The goal of ensuring a meal for every schoolchild in the world seemed a daunting task, even overwhelming, and endlessly complex. Could it really be done, I wondered. And at what cost? Yet, daunting, overwhelming, and complex in comparison with what?

We all know that the nations of the world spend untold sums on military arms and campaigns. Most of these efforts are huge, complex, and driven by ambitious people. Could we not treat the eradication of hunger with the same purpose of effort and commitment? How many meals for school children can be purchased for the price of an aircraft carrier?

I thought about the program George McGovern was championing. Not only was the program admirable, but the goals were attainable. School children around the world could be fed. But to do so would require the political commitment and energy of governments to change their priorities. Compassion would need a higher priority, warmongering less so.

I thought about George McGovern again and wondered if he would have made a great difference to this country if he had been elected president. We would have avoided Watergate and the drumming from office of a president of the United States. And perhaps, most importantly, he might have set us on a course for an even more compassionate foreign policy that looked after the plight of the hungry and suffering in other nations—a course which might have averted our present situation.

For our nation, I would like to see a national policy that promotes contributing more from our bounty to those in need than so aggressively spending our resources to protect the things that we possess. Such a policy, I think, would do more to enhance our image around the world and encourage world peace and good will among nations than does our present military image, which seems to reduce our prestige and encourage violence without promoting peace.

So, while I profess ownership, or legal title to my farm, I have little control over the livestock that I sell. I produce high quality food, while others, some even in my own backyard, go hungry. Under such circumstances, does not my ownership count for very little?

Perhaps the Osage Indians pointed to a better distribution of food. Those who needed food were allowed to have it. An overabundance of food in one group was never permitted, unless an overabundance was available to all groups. The Osage, of course, were not a perfect people—but I suspect they had far fewer problems that we have in our world today.

I would feel better about my little piece of land if I knew that somehow, it was not just tied into the world agricultural economy as it now exists, but an expanded program to help feed hungry people. We have, I think, something to learn from the Osage who controlled the Ozarks for so long. They had a system of distributing food as people needed it. Today people still need food. I would be extremely proud to know that produce from my farm was a part of a plan, a movement, to help feed hungry people around the world and that some

Who Owns My Farm?

meat, grain, or other product from my farm would eventually make its way into school breakfast programs around the world.

I could feel, then, that my farm is not just a commercial venture, but a productive element, part of the interdependent resources of planet earth.

Berwick
Return to the Farm

When I was a small child, lamenting to my grandmother the slow passing of each day and yearning to be a "grown-up," she would respond, "Enjoy every day. When you grow up, time will pass much faster than it does now." I now recognize this response as her own yearning for more of the time that I so bemoaned. I understand what she was saying. I've been farming this piece of land for nearly ten years now. The time has passed ever too quickly.

Sometimes, though, I'm still amazed at how fast I made the decision to actively take part in the operation of the farm. I now realize that my reaction to the opportunity was as spontaneous as the jerk of the knee to the tap of the physician's small rubber mallet. It just could not have been otherwise. At no time did I sit down and ponder over changes that might be wrought in my life by farming this piece of land, nor did I make a long lists of the pros and cons of farming, discuss them with my wife, and then, after much deliberation, allow us to make a joint decision. No, instantaneously, I was cemented to the farm and set to work to turn the land into a productive farm.

The farmstead, amazingly, looked much the same as it had the day our family moved to it almost forty years earlier. The farmhouse, still white in color, silhouetted the skyline, and massive trees, although not the ones present in my youth, continued to surrounded the house. Inside, gas heat had been added, the walls insulated and an upstairs toilet installed. And in the intervening years, additional memories had attached themselves to the house, for this was the place where I had visited my parents many times. It was here, for instance, that, as a young man I brought my new wife, continued to observe

Berwick Return to the Farm

Christmas and other festive holidays, delighted my parents with their first grandchild, saw my parents age and suffered with my father through the chronic disease that eventually took his life. But now, my sister kept a clean and comfortable home inside the house and tended to the duties incumbent on home ownership. The structure was now her home.

Out in the barn, the same "not-readily-observable" transformation had taken place. The huge barn still towered over the barn lot and from the outside looked much the same as it had fifty years before. One could not see by looking from the outside that the old building had required extensive upkeep. At one point after years of being battered by winds and storms, the building had deteriorated to the point of near collapse, and my sister hired a workman to string heavy wire cables from supporting wall to supporting wall to keep the old building upright. Only then was the workman able to climb on the roof and replace the rusting and leaky old roof with new sheets of galvanized metal. But inside, the barn barely resembled the barn of earlier years. Now there were no sacks of feed piled in the corner next to the milking shed. There was no need, for the dairy cows had been sold long ago. The granaries, therefore, were empty of grain, having been turned into storage areas for our precious junk. And, in the milking shed where years before cows raced to their stanchion to be milked, my nephew's sheep and goats lulled throughout the day. The telltale signs of family change, so slow to transpire, were observable everywhere.

Much the same held true for the Berwick community. At first glance it too, greatly resembled the community I had known as a child. The same rough old country roads still tied the community together. Trains, whistling at the crossing below the house, still hurtled down the tracks pulling huge loads of cargo, and Clear Creek still flowed jauntily along the end of the property although with less water. Most of the other old farm houses in the community remained, too, except now they were forty years older. Interspersed among them were newer, larger, more modern homes. The church still stood on the hill, the sanctuary for the spiritual beliefs of the community, although two new additions adorned the structure: Sunday School rooms attached to the rear of the church and an immaculate activities center, complete with kitchen and gymnasium, stood behind the church.

The invisible community changes became obvious one day as I was building fence alongside the road and watching the increased number of passing vehicles—pickup trucks, trucks hauling feed and hay, passenger cars, and occasionally a plodding tractor. Since, as a kid, we knew almost everyone who passed by, I still expected, unconsciously, to recognize the drivers and passengers of these automobiles. But, of course, I knew very few people. Occasionally, an older person that I had known in my childhood would pass and on other occasions friends and boyhood chums, people I had grown up with, saw me and stopped to visit. "We're still youngsters," we would remind ourselves as we became reacquainted, swapping stories about "how things used to be," caught up on the news of the community, and talked about our children and grandchildren.

But mostly those who passed were strangers, coming from homes with which I was not acquainted and going to destinations I was unaware of. No, these were not the same people I had known as a child, nor were they related to people I had known. To me, they were strangers, living in what was becoming increasingly a different community. During the time I had been away there had been a wrenching of the community with which I was unfamiliar.

Berwick Return to the Farm

An uneasiness, a knotting developed in my stomach as I watched these strangers pass by, and for weeks the anxiety returned each time I visited the farm. Eventually I recognized this as my personal rendition of the theme "You can't go home again," the longing so often memorialized in book and song and personal recollection. Was not part of the attraction of returning to the farm the desire to find a way to live my senior years with the same innocence in which I had so thrived during my childhood? Would it be possible to so live among these strangers?

Now, I was uncovering the fact that my purpose for farming was more, much more than having a successful and profitable farm. Surfacing from deep within was a desire, a hope retained from childhood, that I might become a part of the community and neighborhood I had known as a child. Of course, it was obvious that you can't come home again. But, even to become reacquainted with the community, to make new friends and learn about new neighborhood patterns, required a look back at childhood memories. Would I, in the process of becoming a part of this new community in which I had grown up, undermine my childhood memories? The knot in my stomach was fear.

And fear begat fear. Within a few months of beginning to farm, I developed a fear that my decision to farm had been, for all practical purposes, an impractical one. I severely underestimated the time and effort necessary to accomplish nearly everything I wanted to do. I had thought that I could complete the fencing of the farm in two years. But, because of other employment, I could only work on the farm in the winter, on nice days, and even then I was not always available. At the end of two years, only a small portion of the needed fencing was complete. So it was also with the cattle. The herd needed to

Return to the Farm

be doubled in size to maximize their benefit to the farm. But, by doubling the size of the herd I had to keep from market several young heifers who would have provided the income I needed to reinvest in the upkeep of the farm. The needs of the farm were many, the income little, the progress slow.

Unexpectedly, I thought of my father. Although we never spoke of it, had he not faced great disappointment upon moving to the farm with his family and discovering that he could not provide an adequate living for those he loved? Yet, I never heard him say that he wanted to quit the operation, lease, or sell the farm. In fact, quite the opposite happened. The farm remained his passion, even in those tough first years, until such time as he could become a full-time farmer. He never lost that passion.

I think that my father simply tempered his expectations and decided that if he could not make a living immediately, he would be able to do so with the passage of time. It was not until six years after we moved to the farm, when I was a senior in high school, that he doubled the size of the dairy herd and quit working off the farm for good.

My situation, of course, was not nearly as critical as his. Only two people, my wife Karlene and I, comprise our immediate family, and we both have other employment so the lack of income from the farm never threatened our livelihood, as it most certainly threatened our family when I was young. But as did my father, I enjoyed farming—being out of doors, working with cattle and tending to a pecan orchard. Eventually, I thought, we could build a barn, and perhaps even a house. I wanted to succeed. And like him, I tempered my expectations to a longer time frame. It would simply take longer than I had planned to accomplish what I had set out to do.

Thus I continued to work on the farm, but with a new interest in the community of Berwick. As a child I was, by residence, a part of the community. Now, because I worked on the farm, I had again become a part of the community. And, once I decided to continue regardless of the length of time it took, my fears turned again into anticipation and I became curious about the community in a way I had not taken an interest as a child. As a child, for example, it never occurred to me to ask, "Where did the name 'Berwick' come from?" Back then, someone would have probably known the answer.

In fact, it did not occur to me to ask the source of the name until one day when I was looking at a map of Scotland. I discovered there, about an inch below Edinburgh, the town of Berwick-an-Tweed, a Scottish town on the England-Scotland border. The name of that town and our community was spelled the same. Surely there was a connection.

Much of the Ozarks, I knew, was settled by the Scotch, Scotch-Irish, and English. In all probability Berwick had been named by a Scottish settler years ago. But, was it possible to be more exact, perhaps even to pinpoint the time Berwick received its name?

I asked Basil Ferguson, the 80 year-old patriarch of the community who, except for service during World War II, had lived in Berwick all his life. "Why," he said, "it's just always been called Berwick," he responded. He could go no further.

Basil, a descendent of one of the very first families that settled in the Clear Creek Valley, embodied the values and history of the community. Basil related to me how the Ferguson family originally homesteaded first, a few miles south in another valley, close to the then existing Butterfield Stage line. But the family, for reasons now unknown to him, soon moved into the Clear Creek Valley in the 1830s. At that time the rich bottomlands were overgrown with huge trees, and elk, buffalo, deer, and bear roamed the valley. His family, as did many homesteaders, settled near a spring for water, and immediately set to work simultaneously constructing a cabin, clearing the land of trees, and planting crops.

Nor did land at this time go uncontested. Basil remembered family stories of homesteaders settling near springs, only to be forced off the land by other homesteaders asserting their rights of previous claims filed on the property. "How difficult life must have been then," Basil remarked.

Basil knew the location of a vaguely marked Civil War Cemetery of Union soldiers in the area—the number of interred soldiers a mystery. And, he told a story about how railroad workers, building the rail lines throughout the valley in the 1870s, bullied their way across private property without duly compensating the owners until a group of armed landowners, with cocked rifles, finally stopped all construction until railroad officials arrived to offer payment or trade right-of-way for other land which the railroad owned.

I was intrigued that Basil had personal reminiscences of the original town of Berwick. He recounted the businesses in the small town, a train station, general store, a blacksmith shop, grain elevator, and a tomato canning factory. He could also recount the demise of several of the buildings.

Basil had developed into as good a storyteller as his father, Pat, who delighted me with stories in my childhood. At one point he began to tell me a story about how his father had lost three fingers. Basil said that Pat's brother Shock had accidentally cut off the fingers one day when the boys were cutting brush. Basil said that as the two brothers were cutting brush, Pat was supposed to hold tight to a sapling so that Shock could get a clear swing at it. Shock, in an effort to sever the young tree with one blow, took a powerful blow and, before Pat could get out of the way, oops, the errant axe connected with Pat's fingers. Klop, off went three fingers. I recognized this story as the one Pat had told me years earlier, but Pat, undoubtedly had embellished his version. Yet, I must confess, I like Pat's version of the story much better.

We ended our conversation, and my history lesson, by talking about Clear Creek, the purity of the stream in days past, and the pollution Basil had witnessed at various times when raw affluent, released from upstream treatment plants, floated down the creek. We both expressed the hope that the creek still would be flowing in another fifty years.

From our conversation I came to an understanding about the origins of the community, how it had grown over the years, and the forces, both from within and outside the community, that had shaped its character. Change had always been a part of the community, and would, undoubtedly, always continue to do so. I realized as I left Basil's home that, regardless of the origin of the name, the Berwick community preceded our family's arrival in the 1950s and would, in some fashion, exist after the last McGill made an exit from the community.

But, my discussion with Basil shed little new light on the origins of the name Berwick, although now I suspected it must have come at the time of the arrival of the railroad in the 1870s rather than with the very first settlers. Curiosity remained. Was there a way to be certain?

My next stop was the public library at the county seat in Neosho where two very nice ladies, volunteers both, helped me find several books relating to the development of the county. I was amazed when one of the ladies handed me a book about an Edinburgh, Scotland, investment firm, The Missouri Land and Livestock Company, formed in 1882 which, in the 1880s purchased nearly 400,000 acres of land in the southwest corner of Missouri. The group, known locally as the Scotch

Return to the Farm

Company, purchased most of its land from the railroad and sought riches by promoting and then reselling this land to European immigrants or to migrants from the eastern United States. To assure potential settlers that the area would flourish, the Missouri Land and Livestock Company imported fine purebred cattle from Scotland and collected the latest scientific farming information to disseminate to the new settlers. The company, while only modestly successful, retained an office in Neosho on the southwest corner of the square for four decades. I came away from the library convinced that a direct link existed between this company and the naming of the small railroad siding, Berwick.

What a wonderful reason this gave me to book a flight to Edinburgh, Scotland, to search through archives for references to an investment company with holdings in the United States in the 1880s. Would there not be, somewhere, a record of this corporation and minutes of a board meeting enumerating the farms and ranches and towns under the influence of this corporation? What a momentary joy it was, after a day of scrutinizing records in a high-vaulted research room and initiating computer-aided retrieval searches in the most modern of research facilities, to find a reference to the Edinburgh American Land Mortgage Company Limited, registered on September 11, 1878. But alas, none of the names of the directors of this mortgage company matched those of the company in Neosho. I could find no connecting thread.

Nor did I find any connection a few days later when I boarded a train in Edinburgh and traveled to Berwick-an-Tweed, a much fought-over town on the Scotland-Britain border. The countryside around Berwick-an-Tweed, with the Tweed River flowing through it, resembled the Ozark countryside, and I found it easy to imagine an Ozark railway siding named for this town—but the direct connection was not established.

Berwick Return to the Farm

The search for the origin of the name, while to date unsuccessful, changed my perception of this community from before its inception to the present and even beyond. Change is, has, and will always be a part of this valley.

Had I been perceptive as a child, I would have discovered that significant community change was already taking place. Already farms were being sold and traded, usually when older people retired or died, and younger families replaced them. The generation before us had walked down the road to a one-room school. We rode school buses ten miles to a consolidated school. I heard people talk about the time, years before, when the two-story general store at Berwick was literally moved two miles from the dying railroad crossing town to the new Highway 60 that was built for automobiles. I knew that already one feed store in Pierce City, the nearest small town, had permanently closed its doors, a sure sign of dwindling agriculture business.

But when growing up, I was so focused on the present, on baby calves and pigs and being part of a hay crew and playing on the high school basketball team that I did not realize I was among the last generation of people to clearly define Berwick as our home community. I did not see this. I believed Berwick would go on forever in the way I imagined it when I was growing up.

No wonder I had a difficult time recognizing the Berwick community when I returned to the farm.

Slowly, though, a new perception of Berwick, some of it based on the changes of the past fifty years, some of it based on a newly-recognized stability, returned. Most farms, defying national trends, remain small. A few hundred acres is a large farm in the community and several forty- and sixty-acre farms, miniature when compared to the national average, continue to exist. Many of our older neighbors still farm, although now

Return to the Farm

social security and retirement funds supplement their incomes. The drivers of many of the vehicles I see passing I now recognize as owners of small farms who tend a few cattle and head to town every morning to work, attesting to the fact that now Berwick serves as a bedroom community for the towns of the area. On occasion and when needed, a few neighbors still share farm work. And although the 4-H Club ceased many years ago and social gatherings like the sumptuous strawberry feeds are rare, the church remains as both a spiritual and social center for the community. Perhaps the greatest agricultural impact on the community has been the four farms where gigantic broiler houses have been constructed, houses so large that each farm raises a million or more chickens each year—becoming farming factories rather than farming operations, where tons of feed are poured through chickens and net gain per pound of feed (calculated to the fraction of a pound) and death rates of chicks determine profit/loss statements.

Berwick is greatly changed and barely identifiable.

Nor is Berwick alone. Tens of thousands of other rural communities across the nation are making the same transformation—given the loss of farms and farm population and the abilities of large corporations to dominate America's agriculture. In fact, the transformation that is taking place in Berwick is not nearly so great as in most other farming communities. The Ozarks, with the rough terrain of rambling creeks and rolling hillsides, lends itself to small farms and traditional livestock and dairy farms—among the last enterprises to be enveloped by the vertical integration and mass corporate farms. In other communities throughout the United States where the soil is deep and the tillable land stretches for miles, farms are consolidated, the fences removed, and the gigantic tractors tread with ease over thousands of acres in a single plot.

Berwick Return to the Farm

The size of these operations is greater than those of us who work with small farms can conceive.

So, what is there to lament about this change of Berwick and the rest of America's agricultural landscape? Certainly there is the nostalgia, the belief that my childhood home and community was the best possible place for me to have been raised! Don't we all cherish our childhood, our family homes, holiday celebrations, and the innocence and joy of growing up, surrounded by a loving and caring family? Such reflections seem even more memorable when placed in farm and small town settings.

But the loss is much greater than this. I think that the nation's ability to produce food is being greatly diminished. Most who operate small and family farms today hold skills passed from previous generations, older to younger, parents to children. Farmers in each generation have perfected their skills by working the land, raising beef cattle, maintaining dairy herds, and apprising themselves of appropriate research. Their skills come from their experiences of working on a farm. They are, generally, people who are anchored to the land through their lifestyle, and are personally committed to producing an abundance of quality.

What a great contrast this is to the corporate farms that lobby for government subsidies and plant and harvest as dictated by accountants and corporate board members. These huge companies, already international in scope, are as free to import food from abroad as the oil companies they parallel are free to import oil. Decisions concerning America's food supply, a commodity as essential as water and air, become further separated from those that produce the food. America's agriculture has become dependent on the fluttering of the political winds. The result—America's agriculture is in a precarious situation.

Return to the Farm

Given more catastrophes in this country, food, not oil, could become the center of national and international concerns.

I hold more confidence in the ability of small and family farmers to deliver America's food supply than the organizational structures of the international corporations. I believe the staying power of the huge corporations is limited and, like corporations in other sectors of the economy, too large to adjust quickly to adverse economic conditions. In poor economic times, corporations could curtail food production in the face of human hunger. In fact, in this land of great agricultural potential, hunger is already prevalent, while in other countries of the world, it is even more widespread. It takes no great stretch of the imagination to see our mega-farms following in the footsteps of the huge centrally controlled Russian farms—following them, even with the massive government supports they receive, into ineffective producers and suppliers of America's food.

Fortunately, the legacy of productive small and family farms remains in Berwick and in many other agricultural communities around the U.S. where both full and part-time farmers seek better ways, not just to produce, but to market their products. What can be better than fresh produce direct from the farm or a farmer's market?

Models for successful farming operations, other than those imposed by the existing agricultural corporations, exist. Farming enterprises like my father had—older family farms that have struggled from generation to generation and continue to do so—or like the Amish, or some of the new efforts at diversifying agriculture show a determination to maintain an agricultural tradition.

There should be a valuable place in America for those who love to till the soil, and who do it well. The tenacious character of the farming community has not yet nor should it be allowed to die. It is incumbent on us all to see that it so remains.

Suggested Reading

Any of the suggested readings below should challenge the mind to question the values and direction of America's marketplace today.

Wendell Berry is America's foremost thinker and writer concerning land, agriculture, and conservation. Through story, prose, and poetry, Berry documents the unnecessary depletion of our natural resources in order to thrive—at a great cost both to ourselves and following generations. Two of his many books are *The Unsettling of America* and *The Art of the Common Place*.

One should read, or re-read Henry David Thoreau's *Walden* or *Life in the Woods*. This American classic on the virtues of the quiet, self-directed life should be required reading of all.

Gene Logsdon is a farmer and writer whose love of his profession shows through in his books. Two of his most endearing books and articles on farming are *Two Acre Eden* and *Living at Nature's Pace*.

Dedicated publisher/editor/farmer Ron Macher crusades at every opportunity to inform America about small farm living. The centerpiece of his crusade is *The Small Farm Magazine* which carries articles, news briefs, and a variety of relevant Web sites for those seeking information on small farms today. Macher can be reached at 3903 W. Trailridge Rd., Clark, MO 65243-9525; 1-800-633-2535, smallfarmtoday.com or smallfarm@socket.net.

The National Sustainable Agriculture Information Service is the Web site of the ATTRA Project, created and managed by the National Center for Appropriate Technology (NCAT) and funded under a grant from the United States Department of Agriculture's Rural Business-Cooperative Service. A tremendous resource of information/links to almost any aspects of sustainable agriculture. Log on at www.attra.org.

The Amish, a religious and agriculturally based people, present an agricultural alternative to our current bio-engineered and corporate farming practices. The Amish have a lesson for us all. If interested read, John Hostetler's, *Amish Society*.

For readers interested in the Ozarks, *Bittersweet Country* and *Bittersweet Earth* serve as an introduction to the customs, habits, and lifestyles of the early inhabitants of the Ozarks. The books, compiled and edited by Ellen Gray Massey, are taken from interviews by high school students of the tenacious character and life philosophies of "old-timers" in the Ozarks. They are valuable not just for the practices of those interviewed, be it fishing, canoeing, blacksmithing, or raising a garden, but for the character and wisdom shown by those interviewed.

John Ikerd, born on a Missouri farm, has spent a lifetime thinking about, teaching, and researching America's agriculture. He is a devotee of sustainable agriculture. His thoughtful writings can be found in the *Small Farm Today Magazine* and at his Web site at www.ssu.missouri/edu/faculty/jikerd.

The Osage Indians inhabited the Ozark region for 700 years before the first white man appeared. One can only wonder, at the pace of civilization today, if descendants of today's inhabitants will be here so long. For those interested, we suggest John Joseph Mathews, *The Osages, Children of the Middle Waters*; Louis Burns, *A History of the Osage People;* and/or Willard H. Rollings, *The Osage, An Ethnohistorical Study of Hegemony of the Prairie-Plains*

An exciting new trend in marketing farm produce is community-supported agriculture (CSA), a movement comprised of family-owned farms which produce a variety of fresh fruits and vegetables. Neighbors and friends purchase "shares" of produce from a farm, then receive the food as it becomes available throughout the growing season. For more information and a list of participating farms, visit the U.S. Department of Agriculture Web site, www.nal.usda.gov/afsic/csa.

Return to the Farm
by Robert McGill

Are available from:

Your local bookstore

Online at whiteoakpublishing.net
Telephone 1-888-255-8672
Fax 417-272-0451

By Mail:
White Oak Publishing
P.O. Box 188
Reeds Spring, MO 65737

Price per copy $12.95 (softbound, this printing)
Sales tax: MO residents add: 6.475%
Shipping costs: $3.00 first book
             $1.50 each additional book

Order Form (may be copied and mailed or faxed):
Please send copy to:

Name ....

Address ....

City ....State................Zip................

E-mail address (optional): ....

Telephone number (optional): ....

Paid by: Check ❑   Credit card ❑

Visa ❑   Mastercard ❑   Optima ❑   Amex ❑   Disc ❑

Card Number: ....

Name on Card: .... Exp. Date:...........